1 4 Date

Y0-CSA-670

Instant Metric Conversion Tables

Instant Metric Conversion Tables

Domus Books

Originally published in Great Britain
by The Hamlyn Publishing Group Limited,
Feltham, Middlesex

© Copyright 1975 The Hamlyn Publishing Group Limited

First published in the United States in 1975 by
Domus Books
400 Anthony Trail
Northbrook
Illinois 60062

ISBN: 0-89196-001-5

Library of Congress Catalog Card Number: 75-24712

Phototypeset in Great Britain by Tradespools Ltd,
Frome, Somerset
Printed in the United States by
William S. Konecky Associates

Introduction

In the United States the US Customary System, which is
derived from the Imperial System but with significant
differences, has always been the accepted system of
measurement. In 1965 the British government announced
that the metric system of weights and measures, which had
in fact been lawful for most purposes in the United Kingdom
since the end of the nineteenth century, would be adopted
in its entirety. The target date for the substantial completion
of the programme of metrication was 1975 and by that year
considerable progress had been made. In 1971 a report was
made to Congress recommending that the United States
should change to the metric system by means of a
co-ordinated national programme over a period of ten years,
at the end of which the country would be predominantly
metric.

The main object of this book is to provide the means for
making instantaneous conversions from the UK and (where
different) the US systems of weights and measures to their
metric equivalents and vice versa, thus eliminating the time-
consuming effort usually involved in such calculations. These
conversions are set out in the tables on pages 21 to 144.
However the book also includes a detailed synopsis of weights
and measures, which, among other things, provides a useful
guide to the metric system and its relationship to the UK and
US systems.

Contents

Synopsis of Weights and Measures

The International System of Units (SI)

In 1960 the Conférence Générale des Poids et Mesures gave formal approval to the Système International d'Unités (better known under the internationally accepted abbreviation of SI), which is an extension and simplification of the traditional metric system founded in France during the French Revolution. SI, which has now been adopted in most 'metric' countries, is both coherent and comprehensive.

Basic SI Units

Physical Quantity	SI Unit	Symbol
length	metre	m
mass	kilogram	kg
time	second	s
electric current	ampere	A
thermodynamic temperature	kelvin	K
amount of substance	mole	mol
luminous intensity	candela	cd

Some derived SI units with special names

Physical Quantity	SI Unit	Symbol
frequency	hertz	Hz
energy	joule	J
force	newton	N
power	watt	W
electric charge	coulomb	C
potential difference	volt	V
resistance	ohm	Ω
capacitance	farad	F
magnetic flux	weber	Wb
inductance	henry	H
magnetic flux density	telsa	T
luminous flux	lumen	lm
illumination	lux	lx

Supplementary dimensionless units

Physical Quantity	SI Unit	Symbol
plane angle	radian	rad
solid angle	steradian	sr

The multiplying SI prefixes

The multiples and sub-multiples of the basic and derived units are not unrelated to the units as is usual in the UK and US systems (e.g., yard, foot, inch and pound, grain, ounce). They are formed by means of prefixes which are the same, no matter to which unit they are applied.

Prefix	Symbol	Factor by which the unit is multiplied	
tera	T	10^{12} =	1000000000000
giga	G	10^{9} =	1000000000
mega	M	10^{6} =	1000000
kilo	k	10^{3} =	1000
hecto	h	10^{2} =	100
deca	da	10^{1} =	10
deci	d	10^{-1} =	0·1
centi	c	10^{-2} =	0·01
milli	m	10^{-3} =	0·001
micro	μ	10^{-6} =	0·000001
nano	n	10^{-9} =	0·000000001
pico	p	10^{-12} =	0·000000000001
femto	f	10^{-15} =	0·000000000000001
atto	a	10^{-18} =	0·000000000000000001

The Metric System of Weights and Measures

Linear Measure

10 millimetres (mm)	= 1 centimetre (cm)
10 centimetres	= 1 decimetre (dm)
10 decimetres	= 1 metre (m)
10 metres	= 1 decametre (dam)
10 decametres	= 1 hectometre (hm)
10 hectometres	= 1 kilometre (km)

Square Measure

100 square millimetres (mm^2)	= 1 square centimetre (cm^2)
100 square centimetres	= 1 square decimetre (dm^2)
100 square decimetres	= 1 square metre (m^2)
100 square metres	= 1 are (1 square decametre)
100 ares	= 1 hectare (ha) (1 square hectometre)
100 hectares	= 1 square kilometre (km^2)

Cubic Measure

1000 cubic millimetres (mm^3)	= 1 cubic centimetre (cm^3)
1000 cubic centimetres	= 1 cubic decimetre (dm^3)
1000 cubic decimetres	= 1 cubic metre (m^3)

Liquid and Dry Measure

10 millilitres (ml)	= 1 centilitre (cl)
10 centilitres	= 1 decilitre (dl)
10 decilitres	= 1 litre (l)
10 litres	= 1 decalitre (dal)
10 decalitres	= 1 hectolitre (hl)
10 hectolitres	= 1 kilolitre (kl)

The litre, as redefined in 1964 by the Conférence Générale des Poids et Mesures, is equal to one cubic decimetre. The millilitre is equal to one cubic centimetre.

Weight

10 milligrams (mg)	= 1 centigram (cg)
10 centigrams	= 1 decigram (dg)
10 decigrams	= 1 gram (g)
10 grams	= 1 decagram (dag)
10 decagrams	= 1 hectogram (hg)
10 hectograms	= 1 kilogram (kg)
100 kilograms	= 1 quintal (q)
10 quintals	= 1 tonne (t)

The UK and US Systems of Weights and Measures

Linear Measure

12 inches (in)	= 1 foot (ft)
3 feet	= 1 yard (yd)
5½ yards	= 1 rod, pole or perch
4 rods	= 1 chain
10 chains	= 1 furlong
8 furlongs (5280 feet)	= 1 statute mile

The rod, pole or perch is now largely obsolete.

Mariner's Measure

6 feet	= 1 fathom
6080 feet	= 1 nautical mile
6076·12 feet	= 1 international nautical mile

The nautical mile is now being superseded by the international nautical mile.

Surveyor's Measure

7·92 inches	= 1 link
100 links	= 1 chain
80 chains	= 1 mile

Square Measure

144 square inches (in²)	= 1 square foot (ft²)
9 square feet	= 1 square yard (yd²)
30¼ square yards	= 1 square rod, pole or perch
40 square rods	= 1 rood
4 roods (4840 square yards)	= 1 acre (ac)
640 acres	= 1 square mile (mile²)

The square rod, pole or perch and the rood are now largely obsolete

Cubic Measure

1728 cubic inches (in³)	= 1 cubic foot (ft³)
27 cubic feet	= 1 cubic yard (yd³)

UK Liquid Measure

60 minims	= 1 fluid drachm
8 fluid drachms	= 1 fluid ounce (UK fl oz)
5 fluid ounces	= 1 gill
4 gills	= 1 pint (UK pt)
2 pints	= 1 quart
4 quarts	= 1 gallon (UK gal)

The minim and fluid drachm can no longer be used for trade purposes in the United Kingdom.

UK Dry Measure

2 pints	= 1 quart
8 quarts	= 1 peck
4 pecks	= 1 bushel

The peck and bushel can no longer be used for trade purposes in the United Kingdom.

US Liquid Measure

60 minims	= 1 fluid dram
8 fluid drams	= 1 fluid ounce (US fl oz)
4 fluid ounces	= 1 gill
4 gills	= 1 pint (US liq pt)
2 pints	= 1 quart
4 quarts	= 1 gallon (US gal)
42 gallons	= 1 barrel (for petroleum)

US Dry Measure

2 pints	= 1 quart
8 quarts	= 1 peck
4 pecks	= 1 bushel (bu)

Relationship between UK and US units of liquid and dry measure

1 UK minim	= 0·96076 US minim
1 UK fluid drachm	= 0·96076 US fluid dram
1 UK fluid ounce	= 0·96076 US fluid ounce
1 UK gill	= 1·20095 US gill
1 UK pint	= 1·20095 US liquid pint
1 UK quart	= 1·20095 US liquid quart
1 UK gallon (277·42 in³)	= 1·20095 US gallon
1 UK pint	= 1·03206 US dry pint
1 UK quart	= 1·03206 US dry quart
1 UK peck	= 1·03206 US peck
1 UK bushel	= 1·03206 US bushel
1 US minim	= 1·04084 UK minim
1 US fluid dram	= 1·04084 UK fluid drachm
1 US fluid ounce	= 1·04084 UK fluid ounce
1 US gill	= 0·832674 UK gill
1 US liquid pint	= 0·832674 UK pint
1 US liquid quart	= 0·832674 UK quart
1 US gallon (231·00 in³)	= 0·832674 UK gallon
1 US dry pint	= 0·968939 UK pint
1 US dry quart	= 0·968939 UK quart
1 US peck	= 0·968939 UK peck
1 US bushel	= 0·968939 UK bushel

The UK dry pint and dry quart have the same capacity as their counterparts in liquid measure but US dry pints and quarts are not equivalents of US liquid pints and quarts. The UK barrel (used in liquid measures only) varies in capacity between $31\frac{1}{2}$ and 36 UK gallons, depending on what is being measured. As a unit of measurement for fruit and vegetables and dried commodities the US dry barrel is equal to 105 US dry quarts.

The above table is based on BS 350: Part 1. Copies of the complete standard are available from BSI at 2 Park Street, London W1A 2BS.

UK Avoirdupois Weight

27·34 grains (gr)	= 1 dram (dr)
16 drams	= 1 ounce (oz)
16 ounces	= 1 pound (lb)
14 pounds	= 1 stone
2 stones	= 1 quarter
4 quarters	= 1 hundredweight (cwt)
20 hundredweights	= 1 long ton (2240 pounds)
2000 pounds	= 1 short ton (sh ton)

US Avoirdupois Weight

27·34 grains	= 1 dram
16 drams	= 1 ounce
16 ounces	= 1 pound
100 pounds	= 1 short hundredweight (sh cwt)
2000 pounds	= 1 short ton
2240 pounds	= 1 long ton

Troy Weight

24 grains	= 1 pennyweight
20 pennyweights	= 1 ounce
12 ounces	= 1 pound

Apothecaries' Weight

20 grains	= 1 scruple
3 scruples	= 1 drachm
8 drachms	= 1 ounce
12 ounces	= 1 pound

The grain has the same value in the avoirdupois, troy and apothecaries' systems. The ounce troy and the apothecaries' ounce are identical and differ from the avoirdupois ounce. The troy pound has no legal basis in the United Kingdom but is legalized in the United States. The apothecaries' units are now illegal for use in the United Kingdom.

Common Conversion Factors

UK and US to Metric	Metric to UK and US

Linear Measure

1 in = 25·4 mm (exactly)	1 mm = 0·0393701 in
1 in = 2·54 cm (exactly)	1 cm = 0·393701 in
1 ft = 0·3048 m (exactly)	1 m = 3·28084 ft
1 yd = 0·9144 m (exactly)	1 m = 1·09361 yd
1 mile = 1·609344 km (exactly)	1 km = 0·621371 mile

Square Measure

1 in^2 = 6·4516 cm^2 (exactly)	1 cm^2 = 0·1550003 in^2
1 ft^2 = 0·092903 m^2	1 m^2 = 10·7639 ft^2
1 yd^2 = 0·836127 m^2	1 m^2 = 1·19599 yd^2
1 ac = 0·404686 ha	1 ha = 2·47105 ac
1 mile^2 = 2·58999 km^2	1 km^2 = 0·386102 mile^2

Cubic Measure

1 in^3 = 16·387064 cm^3 (exactly)	1 cm^3 = 0·0610237 in^3
1 ft^3 = 0·0283168 m^3	1 m^3 = 35·314725 ft^3
1 yd^3 = 0·764555 m^3	1 m^3 = 1·30795 yd^3

UK Liquid Measure

1 fl oz = 28·4131 ml	1 ml = 0·0351951 fl oz
1 fl oz = 0·2841 dl	1 dl = 3·5195 fl oz
1 pt = 0·568261 l	1 l = 1·75975 pt
1 gal = 4·54609 l	1 l = 0·219969 gal

UK and US to Metric   Metric to UK and US

US Liquid Measure

1 fl oz = 29·5735 ml

1 fl oz = 0·2957 dl

1 pt = 0·473176 l

1 gal = 3·78541 l

1 ml = 0·033814 fl oz

1 dl = 3·3814 fl oz

1 l = 2·11338 pt

1 l = 0·264172 gal

US Dry Measure

1 pt = 0·550610 l

1 bu = 35·2391 l

1 l = 1·81617 pt

1 l = 0·0283776 bu

Avoirdupois Weight

1 gr = 0·0647989 g

1 oz = 28·3495 g

1 lb = 0·453592 kg

1 sh cwt = 0·0453592 t

1 cwt = 0·0508023 t

1 sh ton = 0·907185 t

1 UK ton = 1·01605 t

1 g = 15·4324 gr

1 g = 0·035274 oz

1 kg = 2·20462 lb

1 t = 22·0462 sh cwt

1 t = 19·6841 cwt

1 t = 1·10231 sh ton

1 t = 0·984207 UK ton

Decimals of an inch to millimetres

in	mm	in	mm	in	mm	in	mm
0·001	0·0254	0·026	0·6604	0·051	1·2954	0·076	1·9304
0·002	0·0508	0·027	0·6858	0·052	1·3208	0·077	1·9558
0·003	0·0762	0·028	0·7112	0·053	1·3462	0·078	1·9812
0·004	0·1016	0·029	0·7366	0·054	1·3716	0·079	2·0066
0·005	0·1270	0·030	0·7620	0·055	1·3970	0·080	2·0320
0·006	0·1524	0·031	0·7874	0·056	1·4224	0·081	2·0574
0·007	0·1778	0·032	0·8128	0·057	1·4478	0·082	2·0828
0·008	0·2032	0·033	0·8382	0·058	1·4732	0·083	2·1082
0·009	0·2286	0·034	0·8636	0·059	1·4986	0·084	2·1336
0·010	0·2540	0·035	0·8890	0·060	1·5240	0·085	2·1590
0·011	0·2794	0·036	0·9144	0·061	1·5494	0·086	2·1844
0·012	0·3048	0·037	0·9398	0·062	1·5748	0·087	2·2098
0·013	0·3302	0·038	0·9652	0·063	1·6002	0·088	2·2352
0·014	0·3556	0·039	0·9906	0·064	1·6256	0·089	2·2606
0·015	0·3810	0·040	1·0160	0·065	1·6510	0·090	2·2860
0·016	0·4064	0·041	1·0414	0·066	1·6764	0·091	2·3114
0·017	0·4318	0·042	1·0668	0·067	1·7018	0·092	2·3368
0·018	0·4572	0·043	1·0922	0·068	1·7272	0·093	2·3622
0·019	0·4826	0·044	1·1176	0·069	1·7526	0·094	2·3876
0·020	0·5080	0·045	1·1430	0·070	1·7780	0·095	2·4130
0·021	0·5334	0·046	1·1684	0·071	1·8034	0·096	2·4384
0·022	0·5588	0·047	1·1938	0·072	1·8288	0·097	2·4638
0·023	0·5842	0·048	1·2192	0·073	1·8542	0·098	2·4892
0·024	0·6096	0·049	1·2446	0·074	1·8796	0·099	2·5146
0·025	0·6350	0·050	1·2700	0·075	1·9050	0·100	2·5400

Decimals of an inch to millimetres

in	mm	in	mm	in	mm	in	mm
0·11	2·794	0·36	9·144	0·61	15·494	0·86	21·844
0·12	3·048	0·37	9·398	0·62	15·748	0·87	22·098
0·13	3·302	0·38	9·652	0·63	16·002	0·88	22·352
0·14	3·556	0·39	9·906	0·64	16·256	0·89	22·606
0·15	3·810	0·40	10·160	0·65	16·510	0·90	22·860
0·16	4·064	0·41	10·414	0·66	16·764	0·91	23·114
0·17	4·318	0·42	10·668	0·67	17·018	0·92	23·368
0·18	4·572	0·43	10·922	0·68	17·272	0·93	23·622
0·19	4·826	0·44	11·176	0·69	17·526	0·94	23·876
0·20	5·080	0·45	11·430	0·70	17·780	0·95	24·130
0·21	5·334	0·46	11·684	0·71	18·034	0·96	24·384
0·22	5·588	0·47	11·938	0·72	18·288	0·97	24·638
0·23	5·842	0·48	12·192	0·73	18·542	0·98	24·892
0·24	6·096	0·49	12·446	0·74	18·796	0·99	25·146
0·25	6·350	0·50	12·700	0·75	19·050	1·00	25·400
0·26	6·604	0·51	12·954	0·76	19·304	2·00	50·800
0·27	6·858	0·52	13·208	0·77	19·558	3·00	76·200
0·28	7·112	0·53	13·462	0·78	19·812	4·00	101·600
0·29	7·366	0·54	13·716	0·79	20·066	5·00	127·000
0·30	7·620	0·55	13·970	0·80	20·320	6·00	152·400
0·31	7·874	0·56	14·224	0·81	20·574	7·00	177·800
0·32	8·128	0·57	14·478	0·82	20·828	8·00	203·200
0·33	8·382	0·58	14·732	0·83	21·082	9·00	228·600
0·34	8·636	0·59	14·986	0·84	21·336	10·00	254·000
0·35	8·890	0·60	15·240	0·85	21·590	11·00	279·400

Millimetres to inches

mm	in	mm	in	mm	in	mm	in
1	0·0394	26	1·0236	51	2·0079	76	2·9921
2	0·0787	27	1·0630	52	2·0472	77	3·0315
3	0·1181	28	1·1024	53	2·0866	78	3·0709
4	0·1575	29	1·1417	54	2·1260	79	3·1102
5	0·1969	30	1·1811	55	2·1654	80	3·1496
6	0·2362	31	1·2205	56	2·2047	81	3·1890
7	0·2756	32	1·2598	57	2·2441	82	3·2283
8	0·3150	33	1·2992	58	2·2835	83	3·2677
9	0·3543	34	1·3386	59	2·3228	84	3·3071
10	0·3937	35	1·3780	60	2·3622	85	3·3465
11	0·4331	36	1·4173	61	2·4016	86	3·3858
12	0·4724	37	1·4567	62	2·4409	87	3·4252
13	0·5118	38	1·4961	63	2·4803	88	3·4646
14	0·5512	39	1·5354	64	2·5197	89	3·5039
15	0·5906	40	1·5748	65	2·5591	90	3·5433
16	0·6299	41	1·6142	66	2·5984	91	3·5827
17	0·6693	42	1·6535	67	2·6378	92	3·6221
18	0·7087	43	1·6929	68	2·6772	93	3·6614
19	0·7480	44	1·7323	69	2·7165	94	3·7008
20	0·7874	45	1·7717	70	2·7559	95	3·7402
21	0·8268	46	1·8110	71	2·7953	96	3·7795
22	0·8661	47	1·8504	72	2·8347	97	3·8189
23	0·9055	48	1·8898	73	2·8740	98	3·8583
24	0·9449	49	1·9291	74	2·9134	99	3·8976
25	0·9843	50	1·9685	75	2·9528	100	3·9370

31559

Millimetres to inches

mm	in	mm	in	mm	in	mm	in
110	4·3307	360	14·1732	610	24·0157	860	33·8583
120	4·7244	370	14·5669	620	24·4904	870	34·2520
130	5·1181	380	14·9606	630	24·8031	880	34·6457
140	5·5118	390	15·3543	640	25·1969	890	35·0394
150	5·9055	400	15·7480	650	25·5906	900	35·4331
160	6·2992	410	16·1417	660	25·9843	910	35·8268
170	6·6929	420	16·5354	670	26·3780	920	36·2205
180	7·0866	430	16·9291	680	26·7717	930	36·6142
190	7·4803	440	17·3228	690	27·1654	940	37·0079
200	7·8740	450	17·7165	700	27·5591	950	37·4016
210	8·2677	460	18·1102	710	27·9528	960	37·7953
220	8·6614	470	18·5039	720	28·3465	970	38·1890
230	9·0551	480	18·8976	730	28·7402	980	38·5827
240	9·4488	490	19·2913	740	29·1339	990	38·9764
250	9·8425	500	19·6850	750	29·5276	1000	39·3701
260	10·2362	510	20·0787	760	29·9213		
270	10·6299	520	20·4724	770	30·3150		
280	11·0236	530	20·8661	780	30·7087		
290	11·4173	540	21·2598	790	31·1024		
300	11·8110	550	21·6535	800	31·4961		
310	12·2047	560	22·0472	810	31·8898		
320	12·5984	570	22·4409	820	32·2835		
330	12·9921	580	22·8346	830	32·6772		
340	13·3858	590	23·2283	840	33·0709		
350	13·7795	600	23·6220	850	33·4646		

Fractions of an inch in sixty-fourths to millimetres

in	mm	in	mm	in	mm
1/64	0·397	13/32	10·319	51/64	20·241
1/32	0·794	27/64	10·716	13/16	20·638
3/64	1·191	7/16	11·113	53/64	21·034
1/16	1·587	29/64	11·509	27/32	21·431
5/64	1·984	15/32	11·906	55/64	21·828
3/32	2·381	31/64	12·303	7/8	22·225
7/64	2·778	1/2	12·700	57/64	22·622
1/8	3·175	33/64	13·097	29/32	23·019
9/64	3·572	17/32	13·494	59/64	23·416
5/32	3·969	35/64	13·891	15/16	23·813
11/64	4·366	9/16	14·288	61/64	24·209
3/16	4·763	37/64	14·684	31/32	24·606
13/64	5·159	19/32	15·081	63/64	25·003
7/32	5·556	39/64	15·478	1	25·400
15/64	5·953	5/8	15·875		
1/4	6·350	41/64	16·272		
17/64	6·747	21/32	16·669		
9/32	7·144	43/64	17·066		
19/64	7·541	11/16	17·463		
5/16	7·938	45/64	17·859		
21/64	8·334	23/32	18·256		
11/32	8·731	47/64	18·653		
23/64	9·128	3/4	19·050		
3/8	9·525	49/64	19·447		
25/64	9·922	25/32	19·844		

Inches and fractions of inches in sixteenths to millimetres

in	mm	in	mm	in	mm	in	mm
1 1/16	26·988	2 5/8	66·675	4 3/16	106·363	5 3/4	146·050
1/8	28·575	11/16	68·263	1/4	107·950	13/16	147·638
3/16	30·163	3/4	69·850	5/16	109·538	7/8	149·225
1/4	31·750	13/16	71·438	3/8	111·125	15/16	150·813
5/16	33·338	7/8	73·025	7/16	112·713	6	152·400
3/8	34·925	15/16	74·613	1/2	114·300	1/16	153·988
7/16	36·513	3	76·200	9/16	115·888	1/8	155·575
1/2	38·100	1/16	77·788	5/8	117·475	3/16	157·163
9/16	39·688	1/8	79·375	11/16	119·063	1/4	158·750
5/8	41·275	3/16	80·963	3/4	120·650	5/16	160·338
11/16	42·863	1/4	82·550	13/16	122·238	3/8	161·925
3/4	44·450	5/16	84·138	7/8	123·825	7/16	163·513
13/16	46·038	3/8	85·725	15/16	125·413	1/2	165·100
7/8	47·625	7/16	87·313	5	127·000	9/16	166·688
15/16	49·213	1/2	88·900	1/16	128·588	5/8	168·275
2	50·800	9/16	90·488	1/8	130·175	11/16	169·863
1/16	52·388	5/8	92·075	3/16	131·763	3/4	171·450
1/8	53·975	11/16	93·663	1/4	133·350	13/16	173·038
3/16	55·563	3/4	95·250	5/16	134·938	7/8	174·625
1/4	57·150	13/16	96·838	3/8	136·525	15/16	176·213
5/16	58·738	7/8	98·425	7/16	138·113	7	177·800
3/8	60·325	15/16	100·013	1/2	139·700	1/16	179·388
7/16	61·913	4	101·600	9/16	141·288	1/8	180·975
1/2	63·500	1/16	103·188	5/8	142·875	3/16	182·563
9/16	65·088	1/8	104·775	11/16	144·463	1/4	184·150

Inches and fractions of inches in sixteenths to millimetres

in	mm	in	mm	in	mm
7 5/16	185·738	8 7/8	225·425	10 7/16	265·113
3/8	187·325	15/16	227·013	1/2	266·700
7/16	188·913	9	228·600	9/16	268·288
1/2	190·500	1/16	230·188	5/8	269·875
9/16	192·088	1/8	231·775	11/16	271·463
5/8	193·675	3/16	233·363	3/4	273·050
11/16	195·263	1/4	234·950	13/16	274·638
3/4	196·850	5/16	236·538	7/8	276·225
13/16	198·438	3/8	238·125	15/16	277·813
7/8	200·025	7/16	239·713	11	279·400
15/16	201·613	1/2	241·300	1/16	280·988
8	203·200	9/16	242·888	1/8	282·575
1/16	204·788	5/8	244·475	3/16	284·163
1/8	206·375	11/16	246·063	1/4	285·750
3/16	207·963	3/4	247·650	5/16	287·338
1/4	209·550	13/16	249·238	3/8	288·925
5/16	211·138	7/8	250·825	7/16	290·513
3/8	212·725	15/16	252·413	1/2	292·100
7/16	214·313	10	254·000	9/16	293·688
1/2	215·900	1/16	255·588	5/8	295·275
9/16	217·488	1/8	257·175	11/16	296·863
5/8	219·075	3/16	258·763	3/4	298·450
11/16	220·663	1/4	260·350	13/16	300·038
3/4	222·250	5/16	261·938	7/8	301·625
13/16	223·838	3/8	263·525	15/16	303·213

Feet and inches
to centimetres

ft in	cm	ft in	cm	ft in	cm	ft in	cm
1′ 0″	30·48	3′ 1″	93·98	5′ 2″	157·48	7′ 3″	220·98
1′ 1″	33·02	3′ 2″	96·52	5′ 3″	160·02	7′ 4″	223·52
1′ 2″	35·56	3′ 3″	99·06	5′ 4″	162·56	7′ 5″	226·06
1′ 3″	38·10	3′ 4″	101·60	5′ 5″	165·10	7′ 6″	228·60
1′ 4″	40·64	3′ 5″	104·14	5′ 6″	167·64	7′ 7″	231·14
1′ 5″	43·18	3′ 6″	106·68	5′ 7″	170·18	7′ 8″	233·68
1′ 6″	45·72	3′ 7″	109·22	5′ 8″	172·72	7′ 9″	236·22
1′ 7″	48·26	3′ 8″	111·76	5′ 9″	175·26	7′ 10″	238·76
1′ 8″	50·80	3′ 9″	114·30	5′ 10″	177·80	7′ 11″	241·30
1′ 9″	53·34	3′ 10″	116·84	5′ 11″	180·34	8′ 0″	243·84
1′ 10″	55·88	3′ 11″	119·38	6′ 0″	182·88	8′ 1″	246·38
1′ 11″	58·42	4′ 0″	121·92	6′ 1″	185·42	8′ 2″	248·92
2′ 0″	60·96	4′ 1″	124·46	6′ 2″	187·96	8′ 3″	251·46
2′ 1″	63·50	4′ 2″	127·00	6′ 3″	190·50	8′ 4″	254·00
2′ 2″	66·04	4′ 3″	129·54	6′ 4″	193·04	8′ 5″	256·54
2′ 3″	68·58	4′ 4″	132·08	6′ 5″	195·58	8′ 6″	259·08
2′ 4″	71·12	4′ 5″	134·62	6′ 6″	198·12	8′ 7″	261·62
2′ 5″	73·66	4′ 6″	137·16	6′ 7″	200·66	8′ 8″	264·16
2′ 6″	76·20	4′ 7″	139·70	6′ 8″	203·20	8′ 9″	266·70
2′ 7″	78·74	4′ 8″	142·24	6′ 9″	205·74	8′ 10″	269·24
2′ 8″	81·28	4′ 9″	144·78	6′ 10″	208·28	8′ 11″	271·78
2′ 9″	83·82	4′ 10″	147·32	6′ 11″	210·82	9′ 0″	274·32
2′ 10″	86·36	4′ 11″	149·86	7′ 0″	213·36	9′ 1″	276·86
2′ 11″	88·90	5′ 0″	152·40	7′ 1″	215·90	9′ 2″	279·40
3′ 0″	91·44	5′ 1″	154·94	7′ 2″	218·44	9′ 3″	281·94

Feet and inches to centimetres

ft	in	cm	ft	in	cm	ft	in	cm	ft	in	cm
9'	4"	284·48	11'	5"	347·98	13'	6"	411·48	15'	7"	474·98
9'	5"	287·02	11'	6"	350·52	13'	7"	414·02	15'	8"	477·52
9'	6"	289·56	11'	7"	353·06	13'	8"	416·56	15'	9"	480·06
9'	7"	292·10	11'	8"	355·60	13'	9"	419·10	15'	10"	482·60
9'	8"	294·64	11'	9"	358·14	13'	10"	421·64	15'	11"	485·14
9'	9"	297·18	11'	10"	360·68	13'	11"	424·18	16'	0"	487·68
9'	10"	299·72	11'	11"	363·22	14'	0"	426·72	16'	1"	490·22
9'	11"	302·26	12'	0"	365·76	14'	1"	429·26	16'	2"	492·76
10'	0"	304·80	12'	1"	368·30	14'	2"	431·80	16'	3"	495·30
10'	1"	307·34	12'	2"	370·84	14'	3"	434·34	16'	4"	497·84
10'	2"	309·88	12'	3"	373·38	14'	4"	436·88	16'	5"	500·38
10'	3"	312·42	12'	4"	375·92	14'	5"	439·42	16'	6"	502·92
10'	4"	314·96	12'	5"	378·46	14'	6"	441·96	16'	7"	505·46
10'	5"	317·50	12'	6"	381·00	14'	7"	444·50	16'	8"	508·00
10'	6"	320·04	12'	7"	383·54	14'	8"	447·04	16'	9"	510·54
10'	7"	322·58	12'	8"	386·08	14'	9"	449·58	16'	10"	513·08
10'	8"	325·12	12'	9"	388·62	14'	10"	452·12	16'	11"	515·62
10'	9"	327·66	12'	10"	391·16	14'	11"	454·66	17'	0"	518·16
10'	10"	330·20	12'	11"	393·70	15'	0"	457·20	17'	6"	533·40
10'	11"	332·74	13'	0"	396·24	15'	1"	459·74	18'	0"	548·64
11'	0"	335·28	13'	1"	398·78	15'	2"	462·28	18'	6"	563·88
11'	1"	337·82	13'	2"	401·32	15'	3"	464·82	19'	0"	579·12
11'	2"	340·36	13'	3"	403·86	15'	4"	467·36	19'	6"	594·36
11'	3"	342·90	13'	4"	406·40	15'	5"	469·90	20'	0"	609·60
11'	4"	345·44	13'	5"	408·94	15'	6"	472·44			

Centimetres
to inches

cm	in	cm	in	cm	in	cm	in
1	0·394	26	10·236	51	20·079	76	29·921
2	0·787	27	10·630	52	20·472	77	30·315
3	1·181	28	11·024	53	20·866	78	30·709
4	1·575	29	11·417	54	21·260	79	31·102
5	1·969	30	11·811	55	21·654	80	31·496
6	2·362	31	12·205	56	22·047	81	31·890
7	2·756	32	12·598	57	22·441	82	32·283
8	3·150	33	12·992	58	22·835	83	32·677
9	3·543	34	13·386	59	23·228	84	33·071
10	3·937	35	13·780	60	23·622	85	33·465
11	4·331	36	14·173	61	24·016	86	33·858
12	4·724	37	14·567	62	24·409	87	34·252
13	5·118	38	14·961	63	24·803	88	34·646
14	5·512	39	15·354	64	25·197	89	35·039
15	5·906	40	15·748	65	25·591	90	35·433
16	6·299	41	16·142	66	25·984	91	35·827
17	6·693	42	16·545	67	26·378	92	36·220
18	7·087	43	16·929	68	26·772	93	36·614
19	7·480	44	17·323	69	27·165	94	37·008
20	7·874	45	17·717	70	27·559	95	37·402
21	8·268	46	18·110	71	27·953	96	37·795
22	8·661	47	18·504	72	28·346	97	38·189
23	9·055	48	18·898	73	28·740	98	38·583
24	9·449	49	19·291	74	29·134	99	38·976
25	9·843	50	19·685	75	29·528	100	39·370

cm	in	cm	in	cm	in	cm	in
110	43·307	360	141·732	610	240·157	860	338·583
120	47·244	370	145·699	620	244·094	870	342·520
130	51·181	380	149·606	630	248·031	880	346·457
140	55·118	390	153·543	640	251·969	890	350·394
150	59·055	400	157·480	650	255·906	900	354·331
160	62·992	410	161·417	660	259·843	910	358·268
170	66·929	420	165·354	670	263·780	920	362·205
180	70·866	430	169·291	680	267·717	930	366·142
190	74·803	440	173·228	690	271·654	940	370·079
200	78·740	450	177·165	700	275·591	950	374·016
210	82·677	460	181·102	710	279·528	960	377·953
220	86·614	470	185·039	720	283·465	970	381·890
230	90·551	480	188·976	730	287·402	980	385·827
240	94·488	490	192·913	740	291·339	990	389·764
250	98·425	500	196·850	750	295·276	1000	393·701
260	102·362	510	200·787	760	299·213		
270	106·299	520	204·724	770	303·150		
280	110·236	530	208·661	780	307·087		
290	114·173	540	212·598	790	311·024		
300	118·110	550	216·535	800	314·961		
310	122·047	560	220·472	810	318·898		
320	125·984	570	224·409	820	322·835		
330	129·921	580	228·346	830	326·772		
340	133·858	590	232·283	840	330·709		
350	137·795	600	236·220	850	334·646		

Feet
to metres

ft	m	ft	m	ft	m	ft	m
1	0·3048	26	7·9248	51	15·5448	76	23·1648
2	0·6096	27	8·2296	52	15·8496	77	23·4696
3	0·9144	28	8·5344	53	16·1544	78	23·7744
4	1·2192	29	8·8392	54	16·4592	79	24·0792
5	1·5240	30	9·1440	55	16·7640	80	24·3840
6	1·8288	31	9·4488	56	17·0688	81	24·6888
7	2·1336	32	9·7536	57	17·3736	82	24·9936
8	2·4384	33	10·0584	58	17·6784	83	25·2984
9	2·7432	34	10·3632	59	17·9832	84	25·6032
10	3·0480	35	10·6680	60	18·2880	85	25·9080
11	3·3528	36	10·9728	61	18·5928	86	26·2128
12	3·6576	37	11·2776	62	18·8976	87	26·5176
13	3·9624	38	11·5824	63	19·2024	88	26·8224
14	4·2672	39	11·8872	64	19·5072	89	27·1272
15	4·5720	40	12·1920	65	19·8120	90	27·4320
16	4·8768	41	12·4968	66	20·1168	91	27·7368
17	5·1816	42	12·8016	67	20·4216	92	28·0416
18	5·4864	43	13·1064	68	20·7264	93	28·3464
19	5·7912	44	13·4112	69	21·0312	94	28·6512
20	6·0960	45	13·7160	70	21·3360	95	28·9560
21	6·4008	46	14·0208	71	21·6408	96	29·2608
22	6·7056	47	14·3256	72	21·9456	97	29·5656
23	7·0104	48	14·6304	73	22·2504	98	29·8704
24	7·3152	49	14·9352	74	22·5552	99	30·1752
25	7·6200	50	15·2400	75	22·8600	100	30·4800

Feet
to metres

ft	m	ft	m	ft	m	ft	m
110	33·528	360	109·728	610	185·928	860	262·128
120	36·576	370	112·776	620	188·976	870	265·176
130	39·624	380	115·824	630	192·024	880	268·224
140	42·672	390	118·872	640	195·072	890	271·272
150	45·720	400	121·920	650	198·120	900	274·320
160	48·768	410	124·968	660	201·168	910	277·368
170	51·816	420	128·016	670	204·216	920	280·416
180	54·864	430	131·064	680	207·264	930	283·464
190	57·912	440	134·112	690	210·312	940	286·512
200	60·960	450	137·160	700	213·360	950	289·560
210	64·008	460	140·208	710	216·408	960	292·608
220	67·056	470	143·256	720	219·456	970	295·656
230	70·104	480	146·304	730	222·504	980	298·704
240	73·152	490	149·352	740	225·552	990	301·752
250	76·200	500	152·400	750	228·600	1000	304·800
260	79·248	510	155·448	760	231·648	2000	609·600
270	82·296	520	158·496	770	234·696	3000	914·400
280	85·344	530	161·544	780	237·744	4000	1219·200
290	88·392	540	164·592	790	240·792	5000	1524·000
300	91·440	550	167·640	800	243·840	6000	1828·800
310	94·488	560	170·688	810	246·888	7000	2133·600
320	97·536	570	173·736	820	249·936	8000	2438·400
330	100·584	580	176·784	830	252·984	9000	2743·200
340	103·632	590	179·832	840	256·032	10000	3048·000
350	106·680	600	182·880	850	259·080	25000	7620·000

Metres to feet

m	ft	m	ft	m	ft	m	ft
1	3·281	26	85·302	51	167·323	76	249·344
2	6·562	27	88·583	52	170·604	77	252·625
3	9·843	28	91·864	53	173·885	78	255·906
4	13·123	29	95·144	54	177·165	79	259·186
5	16·404	30	98·425	55	180·446	80	262·467
6	19·685	31	101·706	56	183·727	81	265·748
7	22·966	32	104·987	57	187·008	82	269·029
8	26·247	33	108·267	58	190·289	83	272·310
9	29·528	34	111·549	59	193·570	84	275·591
10	32·808	35	114·829	60	196·850	85	278·871
11	36·089	36	118·110	61	200·131	86	282·152
12	39·370	37	121·391	62	203·412	87	285·433
13	42·651	38	124·672	63	206·693	88	288·714
14	45·932	39	127·953	64	209·974	89	291·995
15	49·213	40	131·234	65	213·255	90	295·276
16	52·493	41	134·514	66	216·535	91	298·556
17	55·774	42	137·795	67	219·816	92	301·837
18	59·055	43	141·076	68	223·097	93	305·118
19	62·336	44	144·357	69	226·378	94	308·399
20	65·617	45	147·638	70	229·659	95	311·680
21	68·898	46	150·919	71	232·940	96	314·961
22	72·179	47	154·199	72	236·221	97	318·242
23	75·459	48	157·480	73	239·501	98	321·522
24	78·740	49	160·761	74	242·782	99	324·803
25	82·021	50	164·042	75	246·063	100	328·084

m	ft	m	ft	m	ft	m	ft
110	360·892	360	1181·102	610	2001·312	860	2821·522
120	393·701	370	1213·911	620	2034·121	870	2854·331
130	426·509	380	1246·719	630	2066·929	880	2887·139
140	459·318	390	1279·528	640	2099·738	890	2919·948
150	492·126	400	1312·336	650	2132·546	900	2952·756
160	524·934	410	1345·144	660	2165·354	910	2985·564
170	557·743	420	1377·953	670	2198·163	920	3018·373
180	590·551	430	1410·761	680	2230·971	930	3051·181
190	623·360	440	1443·570	690	2263·780	940	3083·990
200	656·168	450	1476·378	700	2296·588	950	3116·798
210	688·976	460	1509·186	710	2329·396	960	3149·606
220	721·785	470	1541·995	720	2362·205	970	3182·415
230	754·593	480	1574·803	730	2395·013	980	3215·223
240	787·402	490	1607·612	740	2427·822	990	3248·032
250	820·210	500	1640·420	750	2460·630	1000	3280·840
260	853·018	510	1673·228	760	2493·438	2000	6561·680
270	885·827	520	1706·037	770	2526·247	3000	9842·520
280	918·635	530	1738·845	780	2559·055	4000	13123·600
290	951·444	540	1771·654	790	2591·864	5000	16494·199
300	984·252	550	1804·462	800	2624·672	6000	19685·039
310	1017·060	560	1837·270	810	2657·480	7000	22965·879
320	1049·869	570	1870·079	820	2690·289	8000	26246·719
330	1082·677	580	1902·887	830	2723·097	9000	29527·559
340	1115·486	590	1935·696	840	2755·906	10000	32808·399
350	1148·294	600	1968·504	850	2788·714	25000	82021·000

Yards
to metres

yd	m	yd	m	yd	m	yd	m
1	0·9144	26	23·7744	51	46·6344	76	69·4944
2	1·8288	27	24·6888	52	47·5488	77	70·4088
3	2·7432	28	25·6032	53	48·4632	78	71·3232
4	3·6576	29	26·5176	54	49·3776	79	72·2376
5	4·5720	30	27·4320	55	50·2920	80	73·1520
6	5·4864	31	28·3464	56	51·2064	81	74·0664
7	6·4008	32	29·2608	57	52·1208	82	74·9808
8	7·3152	33	30·1752	58	53·0352	83	75·8952
9	8·2296	34	31·0896	59	53·9496	84	76·8096
10	9·1440	35	32·0040	60	54·8640	85	77·7240
11	10·0584	36	32·9184	61	55·7784	86	78·6384
12	10·9728	37	33·8328	62	56·6928	87	79·5528
13	11·8872	38	34·7472	63	57·6072	88	80·4672
14	12·8016	39	35·6616	64	58·5216	89	81·3816
15	13·7160	40	36·5760	65	59·4360	90	82·2960
16	14·6304	41	37·4904	66	60·3504	91	83·2104
17	15·5448	42	38·4048	67	61·2648	92	84·1248
18	16·4592	43	39·3192	68	62·1792	93	85·0392
19	17·3736	44	40·2336	69	63·0936	94	85·9536
20	18·2880	45	41·1480	70	64·0080	95	86·8680
21	19·2024	46	42·0624	71	64·9224	96	87·7824
22	20·1168	47	42·9768	72	65·8368	97	88·6968
23	21·0312	48	43·8912	73	66·7512	98	89·6112
24	21·9456	49	44·8056	74	67·6656	99	90·5256
25	22·8600	50	45·7200	75	68·5800	100	91·4400

Yards
to metres

yd	m	yd	m	yd	m	yd	m
110	100·584	360	329·184	610	557·784	860	786·384
120	109·728	370	338·328	620	566·928	870	795·528
130	118·872	380	347·472	630	576·072	880	804·672
140	128·016	390	356·616	640	585·216	890	813·816
150	137·160	400	365·760	650	594·360	900	822·960
160	146·304	410	374·904	660	603·504	910	832·104
170	155·448	420	384·048	670	612·648	920	841·248
180	164·592	430	393·192	680	621·792	930	850·392
190	173·736	440	402·336	690	630·936	940	859·536
200	182·880	450	411·480	700	640·080	950	868·680
210	192·024	460	420·624	710	649·224	960	877·824
220	201·168	470	429·768	720	658·368	970	886·968
230	210·312	480	438·912	730	667·512	980	896·112
240	219·456	490	448·056	740	676·656	990	905·256
250	228·600	500	457·200	750	685·800	1000	914·400
260	237·744	510	466·344	760	694·944	2000	1828·800
270	246·888	520	475·488	770	704·088	3000	2743·200
280	256·032	530	484·632	780	713·232	4000	3657·600
290	265·176	540	493·776	790	722·376	5000	4572·000
300	274·320	550	502·920	800	731·520	6000	5486·400
310	283·464	560	512·064	810	740·664	7000	6408·000
320	292·608	570	521·208	820	749·808	8000	7315·200
330	301·752	580	530·352	830	758·952	9000	8229·600
340	310·896	590	539·496	840	768·096	10000	9144·000
350	320·040	600	548·640	850	777·240	25000	22860·000

Metres to yards

m	yd	m	yd	m	yd	m	yd
1	1·0936	26	28·4339	51	55·7743	76	83·1146
2	2·1872	27	29·5276	52	56·8679	77	84·2082
3	3·2808	28	30·6212	53	57·9615	78	85·3018
4	4·3745	29	31·7148	54	59·0551	79	86·3955
5	5·4681	30	32·8084	55	60·1487	80	87·4891
6	6·5617	31	33·9020	56	61·2423	81	88·5827
7	7·6553	32	34·9956	57	62·3360	82	89·6763
8	8·7489	33	36·0892	58	63·4296	83	90·7699
9	9·8425	34	37·1829	59	64·5232	84	91·8635
10	10·9361	35	38·2765	60	65·6168	85	92·9571
11	12·0297	36	39·3701	61	66·7104	86	94·0507
12	13·1234	37	40·4637	62	67·8040	87	95·1444
13	14·2170	38	41·5573	63	68·8976	88	96·2380
14	15·3106	39	42·6509	64	69·9913	89	97·3316
15	16·4042	40	43·7445	65	71·0849	90	98·4252
16	17·4978	41	44·8381	66	72·1785	91	99·5188
17	18·5914	42	45·9318	67	73·2721	92	100·6124
18	19·6850	43	47·0254	68	74·3657	93	101·7060
19	20·7787	44	48·1190	69	75·4593	94	102·7996
20	21·8723	45	49·2126	70	76·5529	95	103·8932
21	22·9659	46	50·3062	71	77·6465	96	104·9868
22	24·0595	47	51·3998	72	78·7402	97	106·0805
23	25·1531	48	52·4934	73	79·8338	98	107·1741
24	26·2467	49	53·5871	74	80·9274	99	108·2677
25	27·3403	50	54·6807	75	82·0210	100	109·3613

m	yd	m	yd	m	yd	m	yd
110	120·2975	360	393·7008	610	667·1044	860	940·5074
120	131·2336	370	404·6363	620	678·0402	870	951·4436
130	142·1697	380	415·5731	630	688·9764	880	962·3797
140	153·1059	390	426·5092	640	699·9125	890	973·3158
150	164·0420	400	437·4453	650	710·8486	900	984·2520
160	174·9781	410	448·3815	660	721·7848	910	995·1881
170	185·9143	420	459·3176	670	732·7209	920	1006·1242
180	196·8504	430	470·2537	680	743·6570	930	1017·0604
190	207·7865	440	481·1898	690	754·5932	940	1027·9965
200	218·7227	450	492·1260	700	765·5293	950	1038·9326
210	229·6588	460	503·0621	710	776·4654	960	1049·8688
220	240·5949	470	513·9982	720	787·4016	970	1060·8049
230	251·5311	480	524·9344	730	798·3377	980	1071·7410
240	262·4672	490	535·8705	740	809·2738	990	1082·6772
250	273·4033	500	546·8066	750	820·2100	1000	1093·6133
260	284·3395	510	557·7428	760	831·1461	2000	2187·2266
270	295·2756	520	568·6789	770	842·0822	3000	3280·8399
280	306·2117	530	579·6150	780	853·0184	4000	4374·4532
290	317·1479	540	590·5512	790	863·9545	5000	5468·0665
300	328·0840	550	601·4873	800	874·8906	6000	6561·6797
310	339·0201	560	612·4234	810	885·8268	7000	7655·2930
320	349·9563	570	623·3596	820	896·7629	8000	8748·9063
330	360·8924	580	634·2957	830	907·6990	9000	9842·5196
340	371·8285	590	645·2318	840	918·6352	10000	10936·1329
350	382·7647	600	656·1680	850	929·5713	25000	27340·3323

Miles
to kilometres

miles	km	miles	km	miles	km	miles	km
1	1·609	26	41·843	51	82·077	76	122·310
2	3·219	27	43·452	52	83·686	77	123·919
3	4·828	28	45·062	53	85·295	78	125·529
4	6·437	29	46·671	54	86·905	79	127·138
5	8·047	30	48·280	55	88·514	80	128·748
6	9·656	31	49·890	56	90·123	81	130·357
7	11·265	32	51·499	57	91·733	82	131·966
8	12·875	33	53·108	58	93·342	83	133·576
9	14·484	34	54·718	59	94·951	84	135·185
10	16·093	35	56·327	60	96·561	85	136·794
11	17·703	36	57·936	61	98·170	86	138·404
12	19·312	37	59·546	62	99·779	87	140·013
13	20·921	38	61·155	63	101·389	88	141·622
14	22·531	39	62·764	64	102·998	89	143·232
15	24·140	40	64·374	65	104·607	90	144·841
16	25·750	41	65·983	66	106·217	91	146·450
17	27·359	42	67·592	67	107·826	92	148·060
18	28·968	43	69·202	68	109·435	93	149·669
19	30·578	44	70·811	69	111·045	94	151·278
20	32·187	45	72·420	70	112·654	95	152·888
21	33·796	46	74·030	71	114·263	96	154·497
22	35·406	47	75·639	72	115·873	97	156·106
23	37·015	48	77·249	73	117·482	98	157·716
24	38·624	49	78·858	74	119·091	99	159·325
25	40·234	50	80·467	75	120·701	100	160·934

Miles
to kilometres

miles	km	miles	km	miles	km	miles	km
110	177·028	360	579·364	610	981·700	860	1384·036
120	193·121	370	595·457	620	997·793	870	1400·129
130	209·215	380	611·551	630	1013·887	880	1416·223
140	225·308	390	627·644	640	1029·980	890	1432·316
150	241·402	400	643·738	650	1046·074	900	1448·410
160	257·495	410	659·831	660	1062·167	910	1464·503
170	273·588	420	675·924	670	1078·260	920	1480·596
180	289·682	430	692·018	680	1094·354	930	1496·690
190	305·775	440	708·111	690	1110·447	940	1512·783
200	321·869	450	724·205	700	1126·541	950	1528·877
210	337·962	460	740·298	710	1142·634	960	1544·970
220	354·056	470	756·392	720	1158·728	970	1561·064
230	370·149	480	772·485	730	1174·821	980	1577·157
240	386·243	490	788·579	740	1190·915	990	1593·251
250	402·336	500	804·672	750	1207·008	1000	1609·344
260	418·429	510	820·765	760	1223·101	2000	3218·688
270	434·523	520	836·859	770	1239·195	3000	4828·032
280	450·616	530	852·952	780	1255·288	4000	6437·376
290	466·710	540	869·046	790	1271·382	5000	8046·720
300	482·803	550	885·139	800	1287·475	6000	9646·064
310	498·897	560	901·233	810	1303·569	7000	11265·408
320	514·990	570	917·326	820	1319·662	8000	12874·752
330	531·084	580	933·420	830	1335·756	9000	14484·096
340	547·177	590	949·513	840	1351·849	10000	16093·440
350	563·270	600	965·606	850	1367·942	25000	40233·600

Kilometres to miles

km	miles	km	miles	km	miles	km	miles
1	0·621	26	16·156	51	31·690	76	47·224
2	1·243	27	16·777	52	32·311	77	47·846
3	1·864	28	17·398	53	32·933	78	48·467
4	2·486	29	18·020	54	33·554	79	49·088
5	3·107	30	18·641	55	34·175	80	49·710
6	3·728	31	19·263	56	34·797	81	50·331
7	4·350	32	19·884	57	35·418	82	50·952
8	4·971	33	20·505	58	36·040	83	51·574
9	5·592	34	21·123	59	36·661	84	52·195
10	6·214	35	21·748	60	37·282	85	52·817
11	6·835	36	22·369	61	37·904	86	53·438
12	7·457	37	22·991	62	38·525	87	54·059
13	8·078	38	23·612	63	39·146	88	54·681
14	8·699	39	24·234	64	39·768	89	55·302
15	9·321	40	24·855	65	40·389	90	55·923
16	9·942	41	25·476	66	41·011	91	56·545
17	10·563	42	26·098	67	41·632	92	57·166
18	11·185	43	26·719	68	42·253	93	57·788
19	11·806	44	27·340	69	42·875	94	58·409
20	12·427	45	27·962	70	43·496	95	59·030
21	13·049	46	28·583	71	44·117	96	59·652
22	13·670	47	29·204	72	44·739	97	60·273
23	14·292	48	29·826	73	45·360	98	60·894
24	14·913	49	30·447	74	45·982	99	61·516
25	15·534	50	31·068	75	46·603	100	62·137

Kilometres
to miles

km	miles	km	miles	km	miles	km	miles
110	68·351	360	223·694	610	379·036	860	534·379
120	74·565	370	229·907	620	385·250	870	540·593
130	80·778	380	236·121	630	391·464	880	546·806
140	86·992	390	242·335	640	397·677	890	553·020
150	93·206	400	248·548	650	403·891	900	559·234
160	99·419	410	254·762	660	410·105	910	565·448
170	105·633	420	260·976	670	416·319	920	571·661
180	111·847	430	267·190	680	422·532	930	577·875
190	118·060	440	273·403	690	428·746	940	584·089
200	124·274	450	279·617	700	434·960	950	590·302
210	130·488	460	285·831	710	441·173	960	596·516
220	136·702	470	292·044	720	447·387	970	602·730
230	142·915	480	298·258	730	453·601	980	608·944
240	149·129	490	304·472	740	459·815	990	615·157
250	155·343	500	310·686	750	466·028	1000	621·371
260	161·556	510	316·899	760	472·242	2000	1242·742
270	167·770	520	323·113	770	478·456	3000	1864·113
280	173·984	530	329·327	780	484·669	4000	2485·484
290	180·198	540	335·540	790	490·883	5000	3106·855
300	186·411	550	341·754	800	497·097	6000	3728·226
310	192·625	560	347·968	810	503·311	7000	4349·597
320	198·839	570	354·181	820	509·524	8000	4970·968
330	205·052	580	360·395	830	515·738	9000	5592·339
340	211·266	590	366·609	840	521·952	10000	6213·710
350	217·480	600	372·823	850	528·165	25000	15534·275

Square inches
to square centimetres

in²	cm²	in²	cm²	in²	cm²	in²	cm²
1	6·4516	26	167·7416	51	329·0316	76	490·3216
2	12·9032	27	174·1932	52	335·4832	77	496·7732
3	19·3548	28	180·6448	53	341·9348	78	503·2248
4	25·8064	29	187·0964	54	348·3864	79	509·6764
5	32·2580	30	193·5480	55	354·8380	80	516·1280
6	38·7096	31	199·9996	56	361·2896	81	522·5796
7	45·1612	32	206·4512	57	367·7412	82	529·0312
8	51·6128	33	212·9028	58	374·1928	83	535·4828
9	58·0644	34	219·3544	59	380·6444	84	541·9344
10	64·5160	35	225·8060	60	387·0960	85	548·3860
11	70·9676	36	232·2576	61	393·5476	86	554·8376
12	77·4192	37	238·7092	62	399·9992	87	561·2892
13	83·8708	38	245·1608	63	406·4508	88	567·7408
14	90·3224	39	251·6124	64	412·9024	89	574·1924
15	96·7740	40	258·0640	65	419·3540	90	580·6440
16	103·2256	41	264·5156	66	425·8056	91	587·0956
17	109·6772	42	270·9672	67	432·2572	92	593·5472
18	116·1288	43	277·4188	68	438·7088	93	599·9988
19	122·5804	44	283·8704	69	445·1604	94	606·4504
20	129·0320	45	290·3220	70	451·6120	95	612·9020
21	135·4836	46	296·7736	71	458·0636	96	619·3536
22	141·9352	47	303·2252	72	464·5152	97	625·8052
23	148·3868	48	309·6768	73	470·9668	98	632·2568
24	154·8384	49	316·1284	74	477·4184	99	638·7084
25	161·2900	50	322·5800	75	483·8700	100	645·1600

Square inches
to square centimetres

in^2	cm^2	in^2	cm^2	in^2	cm^2	in^2	cm^2
110	709·676	360	2322·576	610	3935·476	860	5548·376
120	774·192	370	2387·092	620	3999·992	870	5612·892
130	838·708	380	2451·608	630	4064·508	880	5677·408
140	903·224	390	2516·124	640	4129·024	890	5741·924
150	967·740	400	2580·640	650	4193·540	900	5806·440
160	1032·256	410	2645·156	660	4258·056	910	5870·956
170	1096·772	420	2709·672	670	4322·572	920	5935·472
180	1161·288	430	2774·188	680	4387·088	930	5999·988
190	1225·804	440	2838·704	690	4451·604	940	6064·504
200	1290·320	450	2903·220	700	4516·120	950	6129·020
210	1354·836	460	2967·736	710	4580·636	960	6193·536
220	1419·352	470	3032·252	720	4645·152	970	6258·052
230	1483·868	480	3096·768	730	4709·668	980	6322·568
240	1548·384	490	3161·284	740	4774·184	990	6387·084
250	1612·900	500	3225·800	750	4838·700	1000	6451·600
260	1677·416	510	3290·316	760	4903·216		
270	1741·932	520	3354·832	770	4967·732		
280	1806·448	530	3419·348	780	5032·248		
290	1870·964	540	3483·864	790	5096·764		
300	1935·480	550	3548·380	800	5161·280		
310	1999·996	560	3612·896	810	5225·796		
320	2064·512	570	3677·412	820	5290·312		
330	2129·028	580	3741·928	830	5354·828		
340	2193·544	590	3806·444	840	5419·344		
350	2258·060	600	3870·960	850	5483·860		

Square centimetres to square inches

cm²	in²	cm²	in²	cm²	in²	cm²	in²
1	0·155	26	4·030	51	7·905	76	11·780
2	0·310	27	4·185	52	8·060	77	11·935
3	0·465	28	4·340	53	8·215	78	12·090
4	0·620	29	4·495	54	8·370	79	12·245
5	0·775	30	4·650	55	8·525	80	12·400
6	0·930	31	4·805	56	8·680	81	12·555
7	1·085	32	4·960	57	8·835	82	12·710
8	1·240	33	5·115	58	8·990	83	12·865
9	1·395	34	5·270	59	9·145	84	13·020
10	1·550	35	5·425	60	9·300	85	13·175
11	1·705	36	5·580	61	9·455	86	13·330
12	1·860	37	5·735	62	9·610	87	13·485
13	2·015	38	5·890	63	9·765	88	13·640
14	2·170	39	6·045	64	9·920	89	13·795
15	2·325	40	6·200	65	10·075	90	13·950
16	2·480	41	6·355	66	10·230	91	14·105
17	2·635	42	6·510	67	10·385	92	14·260
18	2·790	43	6·665	68	10·540	93	14·415
19	2·945	44	6·820	69	10·695	94	14·570
20	3·100	45	6·975	70	10·850	95	14·725
21	3·255	46	7·130	71	11·005	96	14·880
22	3·410	47	7·285	72	11·160	97	15·035
23	3·565	48	7·440	73	11·315	98	15·190
24	3·720	49	7·595	74	11·470	99	15·345
25	3·875	50	7·750	75	11·625	100	15·500

cm^2	in^2	cm^2	in^2	cm^2	in^2	cm^2	in^2
110	17·050	360	55·800	610	94·550	860	133·300
120	18·600	370	57·350	620	96·100	870	134·850
130	20·150	380	58·900	630	97·650	880	136·400
140	21·700	390	60·450	640	99·200	890	137·950
150	23·250	400	62·000	650	100·750	900	139·500
160	24·800	410	63·550	660	102·300	910	141·050
170	26·350	420	65·100	670	103·850	920	142·600
180	27·900	430	66·650	680	105·400	930	144·150
190	29·450	440	68·200	690	106·950	940	145·700
200	31·000	450	69·750	700	108·500	950	147·250
210	32·550	460	71·300	710	110·050	960	148·800
220	34·100	470	72·850	720	111·600	970	150·350
230	35·650	480	74·400	730	113·150	980	151·900
240	37·200	490	75·950	740	114·700	990	153·450
250	38·750	500	77·500	750	116·250	1000	155·000
260	40·300	510	79·050	760	117·800	2000	310·000
270	41·850	520	80·600	770	119·350	3000	465·000
280	43·400	530	82·150	780	120·900	4000	620·000
290	44·950	540	83·700	790	122·450	5000	775·000
300	46·500	550	85·250	800	124·000	6000	930·000
310	48·050	560	86·800	810	125·550	7000	1085·000
320	49·600	570	88·350	820	127·100	8000	1240·000
330	51·150	580	89·900	830	128·650	9000	1395·000
340	52·700	590	91·450	840	130·200	10000	1550·000
350	54·250	600	93·000	850	131·750	25000	3875·000

Square feet to square metres

ft²	m²	ft²	m²	ft²	m²	ft²	m²
1	0·0929	26	2·4155	51	4·7381	76	7·0606
2	0·1858	27	2·5084	52	4·8310	77	7·1535
3	0·2787	28	2·6013	53	4·9239	78	7·2464
4	0·3716	29	2·6942	54	5·0168	79	7·3393
5	0·4645	30	2·7871	55	5·1097	80	7·4322
6	0·5574	31	2·8800	56	5·2026	81	7·5251
7	0·6503	32	2·9729	57	5·2955	82	7·6180
8	0·7432	33	3·0658	58	5·3884	83	7·7110
9	0·8361	34	3·1587	59	5·4813	84	7·8039
10	0·9290	35	3·2516	60	5·5742	85	7·8968
11	1·0219	36	3·3445	61	5·6671	86	7·9897
12	1·1148	37	3·4374	62	5·7600	87	8·0826
13	1·2077	38	3·5303	63	5·8529	88	8·1755
14	1·3006	39	3·6232	64	5·9458	89	8·2684
15	1·3935	40	3·7161	65	6·0387	90	8·3613
16	1·4864	41	3·8090	66	6·1316	91	8·4542
17	1·5794	42	3·9019	67	6·2245	92	8·5471
18	1·6723	43	3·9948	68	6·3174	93	8·6400
19	1·7652	44	4·0877	69	6·4103	94	8·7329
20	1·8581	45	4·1806	70	6·5032	95	8·8258
21	1·9510	46	4·2735	71	6·5961	96	8·9187
22	2·0439	47	4·3364	72	6·6890	97	9·0116
23	2·1368	48	4·4593	73	6·7819	98	9·1045
24	2·2297	49	4·5522	74	6·8748	99	9·1974
25	2·3226	50	4·6452	75	6·9677	100	9·2903

Square feet
to square metres

ft^2	m^2	ft^2	m^2	ft^2	m^2	ft^2	m^2
110	10·2193	360	33·4451	610	56·6708	860	79·8966
120	11·1484	370	34·3741	620	57·5999	870	80·8256
130	12·0774	380	35·3031	630	58·5289	880	81·7546
140	13·0064	390	36·2322	640	59·4579	890	82·6837
150	13·9355	400	37·1612	650	60·3870	900	83·6127
160	14·8645	410	38·0902	660	61·3160	910	84·5417
170	15·7935	420	39·0193	670	62·2450	920	85·4708
180	16·7225	430	39·9483	680	63·1740	930	86·3998
190	17·6516	440	40·8773	690	64·1031	940	87·3288
200	18·5806	450	41·8064	700	65·0321	950	88·2579
210	19·5096	460	42·7354	710	65·9611	960	89·1869
220	20·4387	470	43·6644	720	66·8902	970	90·1159
230	21·3677	480	44·5934	730	67·8192	980	91·0449
240	22·2967	490	45·5225	740	68·7482	990	91·9740
250	23·2258	500	46·4515	750	69·6773	1000	92·9030
260	24·1548	510	47·3805	760	70·6063	2000	185·8060
270	25·0838	520	48·3096	770	71·5353	3000	278·7090
280	26·0128	530	49·2386	780	72·4643	4000	371·6120
290	26·9419	540	50·1676	790	73·3934	5000	464·5150
300	27·8709	550	51·0967	800	74·3224	6000	557·4180
310	28·7999	560	52·0257	810	75·2514	7000	650·3210
320	29·7290	570	52·9547	820	76·1805	8000	743·2240
330	30·6580	580	53·8837	830	77·1095	9000	836·1270
340	31·5870	590	54·8128	840	78·0385	10000	929·0300
350	32·5161	600	55·7418	850	78·9676	25000	2322·5750

Square metres to square feet

m^2	ft^2	m^2	ft^2	m^2	ft^2	m^2	ft^2
1	10·764	26	279·862	51	548·959	76	818·057
2	21·528	27	290·626	52	559·723	77	828·821
3	32·292	28	301·389	53	570·487	78	839·585
4	43·056	29	312·153	54	581·251	79	850·349
5	53·820	30	322·917	55	592·015	80	861·113
6	64·583	31	333·681	56	602·779	81	871·877
7	75·347	32	344·445	57	613·543	82	882·641
8	86·111	33	355·209	58	624·307	83	893·405
9	96·875	34	365·973	59	635·071	84	904·168
10	107·639	35	376·737	60	645·835	85	914·932
11	118·403	36	387·501	61	656·599	86	925·696
12	129·167	37	398·265	62	667·362	87	936·460
13	139·931	38	409·029	63	678·126	88	947·224
14	150·695	39	419·793	64	688·890	89	957·988
15	161·459	40	430·556	65	699·654	90	968·752
16	172·223	41	441·320	66	710·418	91	979·516
17	182·986	42	452·084	67	721·182	92	990·280
18	193·750	43	462·848	68	731·946	93	1001·044
19	204·514	44	473·612	69	742·710	94	1011·808
20	215·278	45	484·376	70	753·474	95	1022·571
21	226·042	46	495·140	71	764·238	96	1033·335
22	236·806	47	505·904	72	775·002	97	1044·099
23	247·570	48	516·668	73	785·765	98	1054·863
24	258·334	49	527·432	74	796·529	99	1065·627
25	269·098	50	538·196	75	807·293	100	1076·391

Square metres to square feet

m^2	ft^2	m^2	ft^2	m^2	ft^2	m^2	ft^2
110	1184·030	360	3875·008	610	6565·985	860	9256·96
120	1291·669	370	3982·647	620	6673·624	870	9364·60
130	1399·308	380	4090·286	630	6781·263	880	9472·24
140	1506·947	390	4197·925	640	6888·902	890	9579·88
150	1614·587	400	4305·564	650	6996·542	900	9687·52
160	1722·226	410	4413·203	660	7104·181	910	9795·16
170	1829·865	420	4520·842	670	7211·820	920	9902·80
180	1937·504	430	4628·481	680	7319·459	930	10010·44
190	2045·143	440	4736·120	690	7427·098	940	10118·08
200	2152·782	450	4843·760	700	7534·737	950	10225·72
210	2260·421	460	4951·399	710	7642·376	960	10333·35
220	2368·060	470	5059·038	720	7750·015	970	10440·10
230	2475·699	480	5166·677	730	7857·654	980	10548·63
240	2583·338	490	5274·316	740	7965·293	990	10656·27
250	2690·978	500	5381·955	750	8072·933	1000	10763·91
260	2798·617	510	5489·594	760	8180·572	2000	21527·82
270	2906·256	520	5597·233	770	8288·211	3000	32291·73
280	3013·895	530	5704·872	780	8395·850	4000	43055·64
290	3121·534	540	5812·511	790	8503·489	5000	53819·55
300	3229·173	550	5920·151	800	8611·128	6000	64583·46
310	3336·812	560	6027·790	810	8718·767	7000	75347·37
320	3444·451	570	6135·429	820	8826·406	8000	86111·28
330	3552·090	580	6243·068	830	8934·045	9000	96875·19
340	3659·729	590	6350·707	840	9041·684	10000	107639·10
350	3767·369	600	6458·346	850	9149·324	25000	269097·75

Square yards to square metres

yd²	m²	yd²	m²	yd²	m²	yd²	m²
1	0·8361	26	21·7393	51	42·6425	76	63·5457
2	1·6723	27	22·5754	52	43·4786	77	64·3818
3	2·5084	28	23·4116	53	44·3148	78	65·2179
4	3·3445	29	24·2477	54	45·1509	79	66·0541
5	4·1806	30	25·0838	55	45·9870	80	66·8902
6	5·0168	31	25·9199	56	46·8231	81	67·7263
7	5·8529	32	26·7561	57	47·6593	82	68·5624
8	6·6890	33	27·5922	58	48·4954	83	69·3986
9	7·5251	34	28·4283	59	49·3315	84	70·2347
10	8·3613	35	29·2645	60	50·1676	85	71·0708
11	9·1974	36	30·1006	61	51·0038	86	71·9070
12	10·0335	37	30·9367	62	51·8399	87	72·7431
13	10·8697	38	31·7728	63	52·6760	88	73·5792
14	11·7058	39	32·6090	64	53·5122	89	74·4153
15	12·5419	40	33·4451	65	54·3483	90	75·2515
16	13·3780	41	34·2812	66	55·1844	91	76·0876
17	14·2142	42	35·1173	67	56·0205	92	76·9237
18	15·0503	43	35·9535	68	56·8567	93	77·7598
19	15·8864	44	36·7896	69	57·6928	94	78·5960
20	16·7225	45	37·6257	70	58·5289	95	79·4321
21	17·5587	46	38·4619	71	59·3650	96	80·2682
22	18·3948	47	39·2980	72	60·2012	97	81·1044
23	19·2309	48	40·1341	73	61·0373	98	81·9405
24	20·0671	49	40·9702	74	61·8734	99	82·7766
25	20·9032	50	41·8064	75	62·7096	100	83·6127

Square yards to square metres

yd²	m²	yd²	m²	yd²	m²	yd²	m²
110	91·9740	360	301·0057	610	510·0375	860	719·069
120	100·3352	370	309·3670	620	518·3987	870	727·431
130	108·6965	380	317·7283	630	526·7600	880	735·792
140	117·0578	390	326·0895	640	535·1213	890	744·153
150	125·4191	400	334·4508	650	543·4826	900	752·514
160	133·7803	410	342·8121	660	551·8438	910	760·876
170	142·1416	420	351·1733	670	560·2051	920	769·237
180	150·5029	430	359·5346	680	568·5664	930	777·598
190	158·8641	440	367·8959	690	576·9276	940	785·959
200	167·2254	450	376·2572	700	585·2889	950	794·321
210	175·5867	460	384·6184	710	593·6502	960	802·682
220	183·9479	470	392·9797	720	602·0114	970	811·043
230	192·3092	480	401·3410	730	610·3727	980	819·405
240	200·6705	490	409·7022	740	618·7340	990	827·766
250	209·0318	500	418·0635	750	627·0953	1000	836·127
260	217·3930	510	426·4248	760	635·4565	2000	1672·254
270	225·7543	520	434·7860	770	643·8178	3000	2508·381
280	234·1156	530	443·1473	780	652·1791	4000	3344·508
290	242·4768	540	451·5086	790	660·5403	5000	4180·635
300	250·8381	550	459·8699	800	668·9016	6000	5016·762
310	259·1994	560	468·2311	810	677·2629	7000	5852·889
320	267·5606	570	476·5924	820	685·6241	8000	6689·016
330	275·9219	580	484·9537	830	693·9854	9000	7525·143
340	284·2832	590	493·3149	840	702·3467	10000	8361·270
350	292·6445	600	501·6762	850	710·7080	25000	20903·175

Square metres to square yards

m^2	yd^2	m^2	yd^2	m^2	yd^2	m^2	yd^2
1	1·1960	26	31·0957	51	60·9955	76	90·8952
2	2·3920	27	32·2917	52	62·1915	77	92·0912
3	3·5880	28	33·4877	53	63·3875	78	93·2872
4	4·7840	29	34·6837	54	64·5835	79	94·4832
5	5·9800	30	35·8797	55	65·7795	80	95·6792
6	7·1759	31	37·0757	56	66·9754	81	96·8752
7	8·3719	32	38·2717	57	68·1714	82	98·0712
8	9·5679	33	39·4677	58	69·3674	83	99·2672
9	10·7639	34	40·6637	59	70·5634	84	100·4632
10	11·9599	35	41·8597	60	71·7594	85	101·6592
11	13·1559	36	43·0556	61	72·9554	86	102·8551
12	14·3519	37	44·2516	62	74·1514	87	104·0511
13	15·5479	38	45·4476	63	75·3474	88	105·2471
14	16·7439	39	46·6436	64	76·5434	89	106·4431
15	17·9399	40	47·8396	65	77·7394	90	107·6391
16	19·1358	41	49·0356	66	78·9353	91	108·8351
17	20·3318	42	50·2316	67	80·1313	92	110·0311
18	21·5278	43	51·4276	68	81·3273	93	111·2271
19	22·7238	44	52·6236	69	82·5233	94	112·4231
20	23·9198	45	53·8196	70	83·7193	95	113·6191
21	25·1158	46	55·0155	71	84·9153	96	114·8150
22	26·3118	47	56·2115	72	86·1113	97	116·0110
23	27·5078	48	57·4075	73	87·3073	98	117·2070
24	28·7038	49	58·6035	74	88·5033	99	118·4030
25	29·8998	50	59·7995	75	89·6993	100	119·5990

Square metres to square yards

m^2	yd^2	m^2	yd^2	m^2	yd^2	m^2	yd^2
110	131·5589	360	430·5564	610	729·5539	860	1028·551
120	143·5188	370	442·5163	620	741·5138	870	1040·511
130	155·4787	380	454·4762	630	753·4737	880	1052·471
140	167·4386	390	466·4361	640	765·4336	890	1064·431
150	179·3985	400	478·3960	650	777·3935	900	1076·391
160	191·3584	410	490·3559	660	789·3534	910	1088·351
170	203·3183	420	502·3158	670	801·3133	920	1100·311
180	215·2782	430	514·2757	680	813·2732	930	1112·271
190	227·2381	440	526·2356	690	825·2331	940	1124·231
200	239·1980	450	538·1955	700	837·1930	950	1136·191
210	251·1579	460	550·1554	710	849·1529	960	1148·150
220	263·1178	470	562·1153	720	861·1128	970	1160·110
230	275·0777	480	574·0752	730	873·0727	980	1172·070
240	287·0376	490	586·0351	740	885·0326	990	1184·030
250	298·9975	500	597·9950	750	896·9925	1000	1195·990
260	310·9574	510	609·9549	760	908·9524	2000	2391·980
270	322·9173	520	621·9148	770	920·9123	3000	3587·970
280	334·8772	530	633·8747	780	932·8722	4000	4783·960
290	346·8371	540	645·8346	790	944·8321	5000	5979·950
300	358·7970	550	657·7945	800	956·7920	6000	7175·940
310	370·7569	560	669·7544	810	968·7519	7000	8371·930
320	382·7168	570	681·7143	820	980·7118	8000	9567·920
330	394·6767	580	693·6742	830	992·6717	9000	10763·910
340	406·6366	590	705·6341	840	1004·6316	10000	11959·900
350	418·5965	600	717·5940	850	1016·5915	25000	29899·750

Acres
to hectares

ac	ha	ac	ha	ac	ha	ac	ha
1	0·4047	26	10·5218	51	20·6390	76	30·7561
2	0·8094	27	10·9265	52	21·0437	77	31·1608
3	1·2141	28	11·3312	53	21·4484	78	31·5655
4	1·6187	29	11·7359	54	21·8530	79	31·9702
5	2·0234	30	12·1406	55	22·2577	80	32·3749
6	2·4281	31	12·5453	56	22·6624	81	32·7796
7	2·8328	32	12·9499	57	23·0671	82	33·1843
8	3·2375	33	13·3546	58	23·4718	83	33·5889
9	3·6422	34	13·7593	59	23·8765	84	33·9936
10	4·0469	35	14·1640	60	24·2812	85	34·3983
11	4·4515	36	14·5687	61	24·6858	86	34·8030
12	4·8562	37	14·9734	62	25·0905	87	35·2077
13	5·2609	38	15·3781	63	25·4952	88	35·6124
14	5·6656	39	15·7828	64	25·8999	89	36·0171
15	6·0703	40	16·1874	65	26·3046	90	36·4217
16	6·4750	41	16·5921	66	26·7093	91	36·8264
17	6·8797	42	16·9968	67	27·1140	92	37·2311
18	7·2843	43	17·4015	68	27·5186	93	37·6358
19	7·6890	44	17·8062	69	27·9233	94	38·0405
20	8·0937	45	18·2109	70	28·3280	95	38·4452
21	8·4984	46	18·6156	71	28·7327	96	38·8499
22	8·9031	47	19·0202	72	29·1374	97	39·2545
23	9·3078	48	19·4249	73	29·5421	98	39·6592
24	9·7125	49	19·8296	74	29·9468	99	40·0639
25	10·1171	50	20·2343	75	30·3515	100	40·4686

Acres
to hectares

ac	ha	ac	ha	ac	ha	ac	ha
110	44·515	360	145·687	610	246·858	860	348·030
120	48·562	370	149·734	620	250·905	870	352·077
130	52·609	380	153·781	630	254·952	880	356·124
140	56·656	390	157·828	640	258·999	890	360·171
150	60·703	400	161·874	650	263·046	900	364·217
160	64·750	410	165·921	660	267·093	910	368·264
170	68·797	420	169·968	670	271·140	920	372·311
180	72·843	430	174·015	680	275·186	930	376·358
190	76·890	440	178·062	690	279·233	940	380·405
200	80·937	450	182·109	700	283·280	950	384·452
210	84·984	460	186·156	710	287·327	960	388·499
220	89·031	470	190·202	720	291·374	970	392·545
230	93·078	480	194·249	730	295·421	980	396·592
240	97·125	490	198·296	740	299·468	990	400·639
250	101·172	500	202·343	750	303·515	1000	404·686
260	105·218	510	206·390	760	307·561	2000	809·372
270	109·265	520	210·437	770	311·608	3000	1214·058
280	113·312	530	214·484	780	315·655	4000	1618·744
290	117·359	540	218·530	790	319·702	5000	2023·430
300	121·406	550	222·577	800	323·749	6000	2428·116
310	125·453	560	226·624	810	327·796	7000	2832·802
320	129·500	570	230·671	820	331·843	8000	3237·488
330	133·546	580	234·718	830	335·889	9000	3642·174
340	137·593	590	238·765	840	339·936	10000	4046·860
350	141·640	600	242·812	850	343·983	25000	10117·150

Hectares
to acres

ha	ac	ha	ac	ha	ac	ha	ac
1	2·471	26	64·247	51	126·024	76	187·800
2	4·942	27	66·718	52	128·495	77	190·271
3	7·413	28	69·190	53	130·966	78	192·742
4	9·884	29	71·661	54	133·437	79	195·213
5	12·355	30	74·132	55	135·908	80	197·684
6	14·826	31	76·603	56	138·379	81	200·155
7	17·297	32	79·074	57	140·850	82	202·626
8	19·768	33	81·545	58	143·321	83	205·097
9	22·239	34	84·016	59	145·792	84	207·568
10	24·711	35	86·487	60	148·263	85	210·040
11	27·182	36	88·958	61	150·734	86	212·511
12	29·653	37	91·429	62	153·205	87	214·982
13	32·124	38	93·900	63	155·676	88	217·453
14	34·595	39	96·371	64	158·147	89	219·924
15	37·066	40	98·842	65	160·618	90	222·395
16	39·537	41	101·313	66	163·090	91	224·866
17	42·008	42	103·784	67	165·561	92	227·337
18	44·479	43	106·255	68	168·032	93	229·808
19	46·950	44	108·726	69	170·503	94	232·279
20	49·421	45	111·197	70	172·974	95	234·750
21	51·892	46	113·668	71	175·445	96	237·221
22	54·363	47	116·140	72	177·916	97	239·692
23	56·834	48	118·611	73	180·387	98	242·163
24	59·305	49	121·082	74	182·858	99	244·634
25	61·776	50	123·553	75	185·329	100	247·105

Hectares
to acres

ha	ac	ha	ac	ha	ac	ha	ac
110	271·816	360	889·579	610	1507·342	860	2125·105
120	296·526	370	914·289	620	1532·052	870	2149·815
130	321·237	380	939·000	630	1556·763	880	2174·526
140	345·947	390	963·710	640	1581·473	890	2199·236
150	370·658	400	988·421	650	1606·184	900	2223·947
160	395·368	410	1013·131	660	1630·894	910	2248·657
170	420·079	420	1037·842	670	1655·605	920	2273·368
180	444·789	430	1062·552	680	1680·315	930	2298·078
190	469·500	440	1087·263	690	1705·026	940	2322·789
200	494·210	450	1111·973	700	1729·736	950	2347·499
210	518·921	460	1136·684	710	1754·447	960	2372·210
220	543·631	470	1161·394	720	1779·157	970	2396·920
230	568·342	480	1186·105	730	1803·868	980	2421·631
240	593·052	490	1210·815	740	1828·578	990	2446·341
250	617·763	500	1235·526	750	1853·289	1000	2471·052
260	642·474	510	1260·237	760	1878·000	2000	4942·104
270	667·184	520	1284·947	770	1902·710	3000	7413·156
280	691·895	530	1309·658	780	1927·421	4000	9884·208
290	716·605	540	1334·368	790	1952·131	5000	12355·260
300	741·316	550	1359·079	800	1976·842	6000	14826·312
310	766·026	560	1383·789	810	2001·552	7000	17297·364
320	790·737	570	1408·500	820	2026·263	8000	19768·416
330	815·447	580	1433·210	830	2050·973	9000	22239·468
340	840·158	590	1457·921	840	2075·684	10000	24710·520
350	846·868	600	1482·631	850	2100·394	25000	61776·300

Square miles
to square kilometres

miles²	km²	miles²	km²	miles²	km²	miles²	km²
1	2·590	26	67·340	51	132·089	76	196·839
2	5·180	27	69·930	52	134·679	77	199·429
3	7·770	28	72·520	53	137·269	78	202·019
4	10·360	29	75·110	54	139·859	79	204·609
5	12·950	30	77·700	55	142·449	80	207·199
6	15·540	31	80·290	56	145·039	81	209·789
7	18·130	32	82·880	57	147·629	82	212·379
8	20·720	33	85·470	58	150·219	83	214·969
9	23·310	34	88·060	59	152·809	84	217·559
10	25·900	35	90·650	60	155·399	85	220·149
11	28·490	36	93·240	61	157·989	86	222·739
12	31·080	37	95·830	62	160·579	87	225·329
13	33·670	38	98·420	63	163·169	88	227·919
14	36·260	39	101·010	64	165·759	89	230·509
15	38·850	40	103·600	65	168·349	90	233·099
16	41·440	41	106·190	66	170·939	91	235·689
17	44·030	42	108·780	67	173·529	92	238·279
18	46·620	43	111·369	68	176·119	93	240·869
19	49·210	44	113·959	69	178·709	94	243·459
20	51·800	45	116·549	70	181·299	95	246·049
21	54·390	46	119·139	71	183·889	96	248·639
22	56·980	47	121·729	72	186·479	97	251·229
23	59·570	48	124·319	73	189·069	98	253·819
24	62·160	49	126·909	74	191·659	99	256·409
25	64·750	50	129·499	75	194·249	100	258·999

miles²	km²	miles²	km²	miles²	km²	miles²	km²
110	284·899	360	932·396	610	1579·893	860	2227·390
120	310·799	370	958·296	620	1605·793	870	2253·290
130	336·699	380	984·195	630	1631·693	880	2279·190
140	362·598	390	1010·095	640	1657·592	890	2305·089
150	388·498	400	1035·995	650	1683·492	900	2330·989
160	414·398	410	1061·895	660	1709·392	910	2356·889
170	440·298	420	1087·795	670	1735·292	920	2382·789
180	466·198	430	1113·695	680	1761·192	930	2408·689
190	492·098	440	1139·595	690	1787·092	940	2434·589
200	517·998	450	1165·495	700	1812·992	950	2460·489
210	543·898	460	1191·395	710	1838·892	960	2486·389
220	569·797	470	1217·294	720	1864·791	970	2512·288
230	595·697	480	1243·194	730	1890·691	980	2538·188
240	621·597	490	1269·094	740	1916·591	990	2564·088
250	647·497	500	1294·994	750	1942·491	1000	2589·988
260	673·397	510	1320·894	760	1968·391	2000	5179·980
270	699·297	520	1346·794	770	1994·291	3000	7769·96
280	725·197	530	1372·694	780	2020·191	4000	10359·950
290	751·097	540	1398·594	790	2046·091	5000	12949·940
300	776·996	550	1424·493	800	2071·990	6000	15539·930
310	802·896	560	1450·393	810	2097·890	7000	18129·920
320	828·796	570	1476·293	820	2123·790	8000	20719·900
330	854·696	580	1502·193	830	2149·690	9000	23309·890
340	880·596	590	1528·093	840	2175·590	10000	25899·880
350	906·496	600	1553·993	850	2201·490	25000	64749·700

Square kilometres
to square miles

km²	miles²	km²	miles²	km²	miles²	km²	miles²
1	0·3861	26	10·0387	51	19·6912	76	29·3438
2	0·7722	27	10·4248	52	20·0773	77	29·7299
3	1·1583	28	10·8109	53	20·4634	78	30·1160
4	1·5444	29	11·1970	54	20·8495	79	30·5021
5	1·9305	30	11·5831	55	21·2356	80	30·8882
6	2·3166	31	11·9692	56	21·6217	81	31·2743
7	2·7027	32	12·3553	57	22·0078	82	31·6604
8	3·0888	33	12·7414	58	22·3939	83	32·0465
9	3·4749	34	13·1275	59	22·7800	84	32·4326
10	3·8610	35	13·5136	60	23·1661	85	32·8187
11	4·2471	36	13·8997	61	23·5522	86	33·2048
12	4·6332	37	14·2858	62	23·9383	87	33·5909
13	5·0193	38	14·6719	63	24·3244	88	33·9770
14	5·4054	39	15·0580	64	24·7105	89	34·3631
15	5·7915	40	15·4441	65	25·0966	90	34·7492
16	6·1776	41	15·8302	66	25·4827	91	35·1353
17	6·5637	42	16·2163	67	25·8688	92	35·5214
18	6·9498	43	16·6024	68	26·2549	93	35·9075
19	7·3359	44	16·9885	69	26·6410	94	36·2936
20	7·7220	45	17·3746	70	27·0272	95	36·6797
21	8·1081	46	17·7607	71	27·4133	96	37·0658
22	8·4942	47	18·1468	72	27·7994	97	37·4519
23	8·8803	48	18·5329	73	28·1855	98	37·8380
24	9·2665	49	18·9190	74	28·5716	99	38·2241
25	9·6526	50	19·3051	75	28·9577	100	38·6102

Square kilometres to square miles

km²	miles²	km²	miles²	km²	miles²	km²	miles²
110	42·4712	360	138·9967	610	235·5222	860	332·0477
120	46·3322	370	142·8577	620	239·3832	870	335·9087
130	50·1933	380	146·7188	630	243·2443	880	339·7698
140	54·0543	390	150·5798	640	247·1053	890	343·6308
150	57·9153	400	154·4408	650	250·9663	900	347·4918
160	61·7763	410	158·3018	660	254·8273	910	351·3528
170	65·6373	420	162·1628	670	258·6883	920	355·2138
180	69·4984	430	166·0239	680	262·5494	930	359·0749
190	73·3594	440	169·8849	690	266·4104	940	362·9359
200	77·2204	450	173·7459	700	270·2714	950	366·7969
210	81·0814	460	177·6069	710	274·1324	960	370·6579
220	84·9424	470	181·4679	720	277·9934	970	374·5189
230	88·8035	480	185·3290	730	281·8545	980	378·3800
240	92·6645	490	189·1900	740	285·7155	990	382·2410
250	96·5255	500	193·0510	750	289·5765	1000	386·1020
260	100·3865	510	196·9120	760	293·4375	2000	772·2040
270	104·2475	520	200·7730	770	297·2985	3000	1158·306
280	108·1086	530	204·6341	780	301·1596	4000	1544·408
290	111·9696	540	208·4951	790	305·0206	5000	1930·510
300	115·8306	550	212·3561	800	308·8816	6000	2316·612
310	119·6916	560	216·2171	810	312·7426	7000	2702·714
320	123·5526	570	220·0781	820	316·6036	8000	3088·816
330	127·4137	580	223·9392	830	320·4647	9000	3474·918
340	131·2747	590	227·8002	840	324·3257	10000	3861·020
350	135·1357	600	231·6612	850	328·1867	25000	9652·600

Cubic inches
to cubic centimetres

in^3	cm^3	in^3	cm^3	in^3	cm^3	in^3	cm^3
1	16·387	26	426·064	51	835·740	76	1245·417
2	32·774	27	442·451	52	852·127	77	1261·804
3	49·161	28	458·838	53	868·514	78	1278·191
4	65·548	29	475·225	54	884·901	79	1294·578
5	81·936	30	491·612	55	901·289	80	1310·965
6	98·323	31	508·000	56	917·676	81	1327·352
7	114·710	32	524·386	57	934·063	82	1343·739
8	131·097	33	540·773	58	950·450	83	1360·126
9	147·484	34	557·160	59	966·837	84	1376·513
10	163·871	35	573·547	60	983·224	85	1392·900
11	180·258	36	589·934	61	999·611	86	1409·288
12	196·645	37	606·321	62	1016·998	87	1425·675
13	213·032	38	622·708	63	1032·385	88	1442·062
14	229·419	39	639·096	64	1048·772	89	1458·449
15	245·806	40	655·483	65	1065·159	90	1474·836
16	262·193	41	671·870	66	1081·546	91	1491·223
17	278·580	42	688·257	67	1097·933	92	1507·610
18	294·967	43	704·644	68	1114·320	93	1523·997
19	311·354	44	721·031	69	1130·707	94	1540·384
20	327·741	45	737·418	70	1147·094	95	1556·771
21	344·128	46	753·805	71	1163·482	96	1573·158
22	360·515	47	770·192	72	1179·869	97	1589·545
23	376·903	48	786·579	73	1196·256	98	1605·932
24	393·290	49	802·966	74	1212·643	99	1622·319
25	409·677	50	819·353	75	1229·030	100	1638·706

Cubic inches
to cubic centimetres

in^3	cm^3	in^3	cm^3	in^3	cm^3	in^3	cm^3
110	1802·577	360	5899·343	610	9996·109	860	14092·88
120	1966·448	370	6063·214	620	10159·979	870	14256·75
130	2130·318	380	6227·084	630	10323·850	880	14420·62
140	2294·189	390	6390·955	640	10487·721	890	14584·49
150	2458·060	400	6554·826	650	10651·591	900	14748·36
160	2621·930	410	6718·696	660	10815·462	910	14912·23
170	2785·801	420	6882·567	670	10979·332	920	15076·10
180	2949·672	430	7046·438	680	11143·203	930	15239·97
190	3113·542	440	7210·308	690	11307·074	940	15403·84
200	3277·413	450	7374·179	700	11470·944	950	15567·71
210	3441·283	460	7538·049	710	11634·815	960	15731·58
220	3605·154	470	7701·920	720	11798·686	970	15895·45
230	3769·025	480	7865·791	730	11962·556	980	16059·32
240	3932·895	490	8029·661	740	12126·427	990	16223·19
250	4096·766	500	8193·532	750	12290·298	1000	16387·06
260	4260·637	510	8357·403	760	12454·168	2000	32774·13
270	4424·507	520	8521·273	770	12618·039	3000	49161·19
280	4588·378	530	8685·144	780	12781·910	4000	65548·26
290	4752·249	540	8849·015	790	12945·780	5000	81935·32
300	4916·119	550	9012·885	800	13109·651	6000	98322·38
310	5079·990	560	9176·756	810	13273·521	7000	114709·45
320	5243·860	570	9340·626	820	13437·392	8000	131096·51
330	5407·731	580	9504·497	830	13601·263	9000	147483·58
340	5571·602	590	9668·368	840	13765·133	10000	163870·64
350	5735·472	600	9832·238	850	13929·004	25000	409676·60

Cubic centimetres to cubic inches

cm³	in³	cm³	in³	cm³	in³	cm³	in³
1	0·0610	26	1·5866	51	3·1122	76	4·6378
2	0·1220	27	1·6476	52	3·1732	77	4·6988
3	0·1831	28	1·7087	53	3·2343	78	4·7598
4	0·2441	29	1·7697	54	3·2953	79	4·8209
5	0·3051	30	1·8307	55	3·3563	80	4·8819
6	0·3661	31	1·8917	56	3·4173	81	4·9429
7	0·4272	32	1·9528	57	3·4784	82	5·0039
8	0·4882	33	2·0138	58	3·5394	83	5·0650
9	0·5492	34	2·0748	59	3·6004	84	5·1260
10	0·6102	35	2·1358	60	3·6614	85	5·1870
11	0·6713	36	2·1969	61	3·7224	86	5·2480
12	0·7323	37	2·2579	62	3·7835	87	5·3091
13	0·7933	38	2·3189	63	3·8445	88	5·3701
14	0·8543	39	2·3799	64	3·9055	89	5·4311
15	0·9154	40	2·4409	65	3·9665	90	5·4921
16	0·9764	41	2·5020	66	4·0276	91	5·5532
17	1·0374	42	2·5630	67	4·0886	92	5·6142
18	1·0984	43	2·6240	68	4·1496	93	5·6752
19	1·1595	44	2·6850	69	4·2106	94	5·7362
20	1·2205	45	2·7461	70	4·2717	95	5·7973
21	1·2815	46	2·8071	71	4·3327	96	5·8583
22	1·3425	47	2·8681	72	4·3937	97	5·9193
23	1·4035	48	2·9291	73	4·4547	98	5·9803
24	1·4646	49	2·9902	74	4·5158	99	6·0413
25	1·5256	50	3·0512	75	4·5768	100	6·1024

Cubic centimetres to cubic inches

cm^3	in^3	cm^3	in^3	cm^3	in^3	cm^3	in^3
110	6·7126	360	21·9685	610	37·2245	860	52·4804
120	7·3228	370	22·5788	620	37·8347	870	53·0907
130	7·9331	380	23·1890	630	38·4450	880	53·7009
140	8·5433	390	23·7992	640	39·0552	890	54·3111
150	9·1536	400	24·4095	650	39·6654	900	54·9214
160	9·7638	410	25·0197	660	40·2757	910	55·5316
170	10·3740	420	25·6300	670	40·8859	920	56·1418
180	10·9843	430	26·2402	680	41·4961	930	56·7521
190	11·5945	440	26·8504	690	42·1064	940	57·3623
200	12·2047	450	27·4607	700	42·7166	950	57·9726
210	12·8150	460	28·0709	710	43·3269	960	58·5828
220	13·4252	470	28·6811	720	43·9371	970	59·1930
230	14·0355	480	29·2914	730	44·5473	980	59·8033
240	14·6457	490	29·9016	740	45·1576	990	60·4135
250	15·2559	500	30·5119	750	45·7678	1000	61·0237
260	15·8662	510	31·1221	760	46·3780	2000	122·0474
270	16·4764	520	31·7323	770	46·9883	3000	183·0711
280	17·0866	530	32·3426	780	47·5985	4000	244·0948
290	17·6969	540	32·9528	790	48·2088	5000	305·1185
300	18·3071	550	33·5630	800	48·8190	6000	366·1422
310	18·9174	560	34·1733	810	49·4292	7000	427·1659
320	19·5276	570	34·7835	820	50·0395	8000	488·1896
330	20·1378	580	35·3937	830	50·6497	9000	549·2133
340	20·7481	590	36·0040	840	51·2599	10000	610·2370
350	21·3583	600	36·6142	850	51·8702	25000	1525·5900

Cubic feet
to cubic metres

ft³	m³	ft³	m³	ft³	m³	ft³	m³
1	0·0283	26	0·7362	51	1·4442	76	2·1521
2	0·0566	27	0·7646	52	1·4725	77	2·1804
3	0·0850	28	0·7929	53	1·5008	78	2·2087
4	0·1133	29	0·8212	54	1·5291	79	2·2370
5	0·1416	30	0·8495	55	1·5574	80	2·2654
6	0·1699	31	0·8778	56	1·5857	81	2·2937
7	0·1982	32	0·9061	57	1·6141	82	2·3220
8	0·2265	33	0·9345	58	1·6424	83	2·3503
9	0·2549	34	0·9628	59	1·6707	84	2·3786
10	0·2832	35	0·9911	60	1·6990	85	2·4069
11	0·3115	36	1·0194	61	1·7273	86	2·4353
12	0·3398	37	1·0477	62	1·7556	87	2·4636
13	0·3681	38	1·0760	63	1·7840	88	2·4919
14	0·3964	39	1·1044	64	1·8123	89	2·5202
15	0·4248	40	1·1327	65	1·8406	90	2·5485
16	0·4531	41	1·1610	66	1·8689	91	2·5768
17	0·4814	42	1·1893	67	1·8972	92	2·6052
18	0·5097	43	1·2176	68	1·9256	93	2·6335
19	0·5380	44	1·2459	69	1·9539	94	2·6618
20	0·5663	45	1·2743	70	1·9822	95	2·6901
21	0·5947	46	1·3026	71	2·0105	96	2·7184
22	0·6230	47	1·3309	72	2·0388	97	2·7467
23	0·6513	48	1·3592	73	2·0671	98	2·7751
24	0·6796	49	1·3875	74	2·0955	99	2·8034
25	0·7079	50	1·4158	75	2·1238	100	2·8317

Cubic feet
to cubic metres

ft³	m³	ft³	m³	ft³	m³	ft³	m³
110	3·1149	360	10·1941	610	17·2733	860	24·3525
120	3·3980	370	10·4772	620	17·5564	870	24·6357
130	3·6812	380	10·7604	630	17·8396	880	24·9188
140	3·9644	390	11·0436	640	18·1228	890	25·2020
150	4·2475	400	11·3267	650	18·4060	900	25·4852
160	4·5307	410	11·6099	660	18·6891	910	25·7683
170	4·8139	420	11·8931	670	18·9723	920	26·0515
180	5·0970	430	12·1762	680	19·2555	930	26·3347
190	5·3802	440	12·4594	690	19·5386	940	26·6178
200	5·6634	450	12·7426	700	19·8218	950	26·9010
210	5·9465	460	13·0257	710	20·1050	960	27·1842
220	6·2297	470	13·3089	720	20·3881	970	27·4673
230	6·5129	480	13·5921	730	20·6713	980	27·7505
240	6·7960	490	13·8753	740	20·9545	990	28·0337
250	7·0792	500	14·1584	750	21·2376	1000	28·3168
260	7·3624	510	14·4416	760	21·5208	2000	56·6337
270	7·6456	520	14·7248	770	21·8040	3000	84·9505
280	7·9287	530	15·0079	780	22·0871	4000	113·2670
290	8·2119	540	15·2911	790	22·3703	5000	141·5840
300	8·4951	550	15·5743	800	22·6535	6000	169·9010
310	8·7782	560	15·8574	810	22·9366	7000	198·2180
320	9·0614	570	16·1406	820	23·2198	8000	226·5350
330	9·3446	580	16·4238	830	23·5030	9000	254·8520
340	9·6277	590	16·7069	840	23·7862	10000	283·1680
350	9·9109	600	16·9901	850	24·0693	25000	707·9210

Cubic metres to cubic feet

m^3	ft^3	m^3	ft^3	m^3	ft^3	m^3	ft^3
1	35·315	26	918·181	51	1801·048	76	2683·915
2	70·629	27	953·496	52	1836·363	77	2719·229
3	105·944	28	988·811	53	1871·677	78	2754·544
4	141·259	29	1024·125	54	1906·992	79	2789·859
5	176·573	30	1059·440	55	1942·307	80	2825·173
6	211·888	31	1094·755	56	1977·621	81	2860·488
7	247·203	32	1130·069	57	2012·936	82	2895·803
8	282·517	33	1165·384	58	2048·251	83	2931·117
9	317·832	34	1200·699	59	2083·565	84	2966·432
10	353·147	35	1236·013	60	2118·880	85	3001·747
11	388·461	36	1271·328	61	2154·195	86	3037·061
12	423·776	37	1306·643	62	2189·509	87	3072·376
13	459·091	38	1341·957	63	2224·824	88	3107·691
14	494·405	39	1377·272	64	2260·139	89	3143·005
15	529·720	40	1412·587	65	2295·453	90	3178·320
16	565·035	41	1447·901	66	2330·768	91	3213·635
17	600·349	42	1483·216	67	2366·083	92	3248·949
18	635·664	43	1518·531	68	2401·397	93	3284·264
19	670·979	44	1553·845	69	2436·712	94	3319·579
20	706·293	45	1589·160	70	2472·027	95	3354·893
21	741·608	46	1624·475	71	2507·341	96	3390·208
22	776·923	47	1659·789	72	2542·656	97	3425·523
23	812·237	48	1695·104	73	2577·971	98	3460·837
24	847·552	49	1730·419	74	2613·285	99	3496·152
25	882·867	50	1765·733	75	2648·600	100	3531·467

Cubic metres
to cubic feet

m^3	ft^3	m^3	ft^3	m^3	ft^3	m^3	ft^3
110	3884·613	360	12713·280	610	21541·947	860	30370·61
120	4237·760	370	13066·427	620	21895·093	870	30723·76
130	4590·907	380	13419·573	630	22248·240	880	31076·91
140	4944·053	390	13772·720	640	22601·387	890	31430·05
150	5297·200	400	14125·867	650	22954·533	900	31783·20
160	5650·347	410	14479·013	660	23307·680	910	32136·35
170	6003·493	420	14832·160	670	23660·827	920	32489·49
180	6356·640	430	15185·307	680	24013·973	930	32842·64
190	6709·787	440	15538·453	690	24367·120	940	33195·79
200	7062·933	450	15891·600	700	24720·267	950	33548·93
210	7416·080	460	16244·747	710	25073·413	960	33902·08
220	7769·227	470	16597·893	720	25426·560	970	34255·23
230	8122·373	480	16951·040	730	25779·707	980	34608·37
240	8475·520	490	17304·187	740	26132·853	990	34961·52
250	8828·667	500	17657·333	750	26486·000	1000	35314·67
260	9181·813	510	18010·480	760	26839·147	2000	70629·33
270	9534·960	520	18363·627	770	27192·293	3000	105944·00
280	9888·107	530	18716·773	780	27545·440	4000	141258·67
290	10241·253	540	19069·920	790	27898·587	5000	176573·33
300	10594·400	550	19423·067	800	28251·733	6000	211888·00
310	10947·547	560	19776·213	810	28604·880	7000	247202·67
320	11300·693	570	20129·360	820	28958·027	8000	282517·33
330	11653·840	580	20482·507	830	29311·173	9000	317832·00
340	12006·987	590	20835·653	840	29664·320	10000	353146·67
350	12360·133	600	21188·800	850	30017·467	25000	882866·70

Cubic yards to cubic metres

yd³	m³	yd³	m³	yd³	m³	yd³	m³
1	0·7646	26	19·8784	51	38·9923	76	58·1062
2	1·5291	27	20·6430	52	39·7569	77	58·8707
3	2·2937	28	21·4075	53	40·5214	78	59·6353
4	3·0582	29	22·1721	54	41·2860	79	60·3998
5	3·8228	30	22·9366	55	42·0505	80	61·1644
6	4·5873	31	23·7012	56	42·8151	81	61·9289
7	5·3519	32	24·4658	57	43·5796	82	62·6935
8	6·1164	33	25·2303	58	44·3442	83	63·4581
9	6·8810	34	25·9949	59	45·1087	84	64·2226
10	7·6455	35	26·6594	60	45·8733	85	64·9872
11	8·4101	36	27·5240	61	46·6378	86	65·7517
12	9·1747	37	28·2885	62	47·4024	87	66·5163
13	9·9392	38	29·0531	63	48·1670	88	67·2808
14	10·7038	39	29·8176	64	48·9315	89	68·0454
15	11·4683	40	30·5822	65	49·6961	90	68·8099
16	12·2329	41	31·3467	66	50·4606	91	69·5745
17	12·9974	42	32·1113	67	51·2252	92	70·3390
18	13·7620	43	32·8759	68	51·9897	93	71·1036
19	14·5265	44	33·6404	69	52·7543	94	71·8682
20	15·2911	45	34·4050	70	53·5188	95	72·6327
21	16·0557	46	35·1695	71	54·2834	96	73·3973
22	16·8202	47	35·9341	72	55·0480	97	74·1618
23	17·5848	48	36·6986	73	55·8125	98	74·9264
24	18·3493	49	37·4632	74	56·5771	99	75·6909
25	19·1139	50	38·2277	75	57·3416	100	76·4555

Cubic yards
to cubic metres

yd^3	m^3	yd^3	m^3	yd^3	m^3	yd^3	m^3
110	84·1011	360	275·2398	610	466·3786	860	657·517
120	91·7466	370	282·8854	620	474·0241	870	665·163
130	99·3922	380	290·5309	630	481·6697	880	672·808
140	107·0377	390	298·1765	640	489·3152	890	680·454
150	114·6833	400	305·8220	650	496·9608	900	688·100
160	122·3288	410	313·4676	660	504·6063	910	695·745
170	129·9744	420	321·1131	670	512·2519	920	703·391
180	137·6199	430	328·7587	680	519·8974	930	711·036
190	145·2655	440	336·4042	690	527·5430	940	718·682
200	152·9110	450	344·0498	700	535·1885	950	726·327
210	160·5566	460	351·6953	710	542·8341	960	733·973
220	168·2021	470	359·3409	720	550·4796	970	741·618
230	175·8477	480	366·9864	730	558·1252	980	749·264
240	183·4932	490	374·6320	740	565·7707	990	756·909
250	191·1388	500	382·2775	750	573·4163	1000	764·555
260	198·7843	510	389·9231	760	581·0618	2000	1529·110
270	206·4299	520	397·5686	770	588·7074	3000	2293·665
280	214·0754	530	405·2142	780	596·3529	4000	3058·220
290	221·7210	540	412·8597	790	603·9985	5000	3822·775
300	229·3665	550	420·5053	800	611·6440	6000	4587·330
310	237·0121	560	428·1508	810	619·2896	7000	5351·885
320	244·6576	570	435·7964	820	626·9351	8000	6116·440
330	252·3032	580	443·4419	830	634·5807	9000	6880·995
340	259·9487	590	451·0875	840	642·2262	10000	7645·550
350	267·5943	600	458·7330	850	649·8718	25000	19113·875

Cubic metres
to cubic yards

m^3	yd^3	m^3	yd^3	m^3	yd^3	m^3	yd^3
1	1·3080	26	34·0067	51	66·7055	76	99·4042
2	2·6159	27	35·3147	52	68·0134	77	100·7122
3	3·9239	28	36·6226	53	69·3214	78	102·0201
4	5·2318	29	37·9306	54	70·6293	79	103·3281
5	6·5398	30	39·2385	55	71·9373	80	104·6360
6	7·8477	31	40·5465	56	73·2452	81	105·9440
7	9·1557	32	41·8544	57	74·5532	82	107·2519
8	10·4636	33	43·1624	58	75·8611	83	108·5599
9	11·7716	34	44·4703	59	77·1691	84	109·8678
10	13·0795	35	45·7783	60	78·4770	85	111·1758
11	14·3875	36	47·0862	61	79·7850	86	112·4837
12	15·6954	37	48·3942	62	81·0929	87	113·7917
13	17·0034	38	49·7021	63	82·4009	88	115·0996
14	18·3113	39	51·0101	64	83·7088	89	116·4076
15	19·6193	40	52·3180	65	85·0168	90	117·7155
16	20·9272	41	53·6260	66	86·3247	91	119·0235
17	22·2352	42	54·9339	67	87·6327	92	120·3314
18	23·5431	43	56·2419	68	88·9406	93	121·6394
19	24·8511	44	57·5498	69	90·2486	94	122·9473
20	26·1590	45	58·8578	70	91·5565	95	124·2553
21	27·4670	46	60·1657	71	92·8645	96	125·5632
22	28·7749	47	61·4737	72	94·1724	97	126·8712
23	30·0829	48	62·7816	73	95·4804	98	128·1791
24	31·3908	49	64·0896	74	96·7883	99	129·4871
25	32·6988	50	65·3975	75	98·0963	100	130·7950

m^3	yd^3	m^3	yd^3	m^3	yd^3	m^3	yd^3
110	143·8745	360	470·8620	610	797·8495	860	1124·837
120	156·9540	370	483·9415	620	810·9290	870	1137·917
130	170·0335	380	497·0210	630	824·0085	880	1150·996
140	183·1130	390	510·1005	640	837·0880	890	1164·076
150	196·1925	400	523·1800	650	850·1675	900	1177·155
160	209·2720	410	536·2595	660	863·2470	910	1190·235
170	222·3515	420	549·3390	670	876·3265	920	1203·314
180	235·4310	430	562·4185	680	889·4060	930	1216·394
190	248·5105	440	575·4980	690	902·4855	940	1229·473
200	261·5900	450	588·5775	700	915·5650	950	1242·553
210	274·6695	460	601·6570	710	928·6445	960	1255·632
220	287·7490	470	614·7365	720	941·7240	970	1268·712
230	300·8285	480	627·8160	730	954·8035	980	1281·791
240	313·9080	490	640·8955	740	967·8830	990	1294·871
250	326·9875	500	653·9750	750	980·9625	1000	1307·950
260	340·0670	510	667·0545	760	994·0420	2000	2615·900
270	353·1465	520	680·1340	770	1007·1215	3000	3923·850
280	366·2260	530	693·2135	780	1020·2010	4000	5231·800
290	379·3055	540	706·2930	790	1033·2805	5000	6539·750
300	392·3850	550	719·3725	800	1046·3600	6000	7847·700
310	405·4645	560	732·4520	810	1059·4395	7000	9155·650
320	418·5440	570	745·5315	820	1072·5190	8000	10463·600
330	431·6235	580	758·6110	830	1085·5985	9000	11771·550
340	444·7030	590	771·6905	840	1098·6780	10000	13079·500
350	457·7825	600	784·7700	850	1111·7575	25000	32698·750

UK fluid ounces to millilitres

UK fl oz	ml	UK fl oz	ml	UK fl oz	ml	UK fl oz	ml
1	28·41	26	738·74	51	1449·07	76	2159·39
2	56·83	27	767·15	52	1477·48	77	2187·81
3	85·24	28	795·57	53	1505·89	78	2216·22
4	113·65	29	823·98	54	1534·31	79	2244·63
5	142·07	30	852·39	55	1562·72	80	2273·05
6	170·48	31	880·81	56	1591·13	81	2301·46
7	198·89	32	909·23	57	1619·54	82	2329·87
8	227·31	33	937·63	58	1647·96	83	2358·28
9	255·72	34	966·04	59	1676·37	84	2386·70
10	284·13	35	994·46	60	1704·78	85	2415·11
11	312·54	36	1022·87	61	1733·20	86	2443·52
12	340·96	37	1051·28	62	1761·61	87	2471·94
13	369·37	38	1079·70	63	1790·02	88	2500·35
14	397·78	39	1108·11	64	1818·44	89	2528·76
15	426·20	40	1136·52	65	1846·85	90	2557·18
16	454·61	41	1164·94	66	1875·26	91	2585·59
17	483·02	42	1193·35	67	1903·68	92	2614·00
18	511·44	43	1221·76	68	1932·09	93	2642·41
19	539·85	44	1250·17	69	1960·50	94	2670·83
20	568·26	45	1278·59	70	1988·91	95	2699·24
21	596·67	46	1307·00	71	2017·33	96	2727·65
22	625·09	47	1335·41	72	2045·74	97	2756·07
23	653·50	48	1363·83	73	2074·15	98	2784·48
24	681·91	49	1392·24	74	2102·57	99	2812·89
25	710·33	50	1420·65	75	2130·98	100	2841·31

Millilitres
to UK fluid ounces

ml	UK fl oz	ml	UK fl oz	ml	UK fl oz	ml	UK fl oz
10	0·3520	260	9·1507	510	17·9495	760	26·7482
20	0·7039	270	9·5027	520	18·3014	770	27·1002
30	1·0559	280	9·8546	530	18·6534	780	27·4521
40	1·4078	290	10·2066	540	19·0053	790	27·8041
50	1·7598	300	10·5585	550	19·3573	800	28·1560
60	2·1117	310	10·9105	560	19·7092	810	28·5080
70	2·4637	320	11·2624	570	20·0612	820	28·8599
80	2·8156	330	11·6144	580	20·4131	830	29·2119
90	3·1676	340	11·9663	590	20·7651	840	29·5638
100	3·5195	350	12·3183	600	21·1170	850	29·9158
110	3·8715	360	12·6702	610	21·4690	860	30·2677
120	4·2234	370	13·0222	620	21·8209	870	30·6197
130	4·5754	380	13·3741	630	22·1729	880	30·9716
140	4·9273	390	13·7261	640	22·5248	890	31·3236
150	5·2793	400	14·0780	650	22·8768	900	31·6755
160	5·6312	410	14·4300	660	23·2287	910	32·0275
170	5·9832	420	14·7819	670	23·5807	920	32·3794
180	6·3351	430	15·1339	680	23·9326	930	32·7314
190	6·6871	440	15·4858	690	24·2846	940	33·0833
200	7·0390	450	15·8378	700	24·6365	950	33·4353
210	7·3910	460	16·1897	710	24·9885	960	33·7872
220	7·7429	470	16·5417	720	25·3404	970	34·1392
230	8·0949	480	16·8936	730	25·6924	980	34·4911
240	8·4468	490	17·2456	740	26·0443	990	34·8431
250	8·7988	500	17·5975	750	26·3963	1000	35·1950

US fluid ounces to millilitres

US fl oz	ml	US fl oz	ml	US fl oz	ml	US fl oz	ml
1	29·57	26	768·91	51	1508·25	76	2247·59
2	59·15	27	798·48	52	1537·82	77	2277·16
3	88·72	28	828·06	53	1567·40	78	2306·73
4	118·29	29	857·63	54	1596·97	79	2336·31
5	147·87	30	887·21	55	1626·54	80	2365·88
6	177·44	31	916·78	56	1656·12	81	2395·45
7	207·01	32	946·35	57	1685·69	82	2425·03
8	236·59	33	975·93	58	1715·26	83	2454·60
9	266·16	34	1005·50	59	1744·84	84	2484·17
10	295·74	35	1035·07	60	1774·41	85	2513·75
11	325·31	36	1064·65	61	1803·98	86	2543·32
12	354·88	37	1094·22	62	1833·56	87	2572·89
13	384·46	38	1123·79	63	1863·13	88	2602·47
14	414·03	39	1153·37	64	1892·70	89	2632·04
15	443·60	40	1182·94	65	1922·28	90	2661·62
16	473·18	41	1212·51	66	1951·85	91	2691·19
17	502·75	42	1242·09	67	1981·42	92	2720·76
18	532·32	43	1271·66	68	2011·00	93	2750·34
19	561·90	44	1301·23	69	2040·57	94	2779·91
20	591·47	45	1330·81	70	2070·15	95	2809·48
21	621·04	46	1360·38	71	2099·72	96	2839·06
22	650·62	47	1389·95	72	2129·29	97	2868·63
23	680·19	48	1419·53	73	2158·87	98	2898·20
24	709·76	49	1449·10	74	2188·44	99	2927·78
25	739·34	50	1478·68	75	2218·01	100	2957·35

Millilitres
to US fluid ounces

ml	US fl oz	ml	US fl oz	ml	US fl oz	ml	US fl oz
10	0·3381	260	8·7916	510	17·2451	760	25·6986
20	0·6763	270	9·1298	520	17·5833	770	26·0368
30	1·0144	280	9·4679	530	17·9214	780	26·3749
40	1·3526	290	9·8061	540	18·2596	790	26·7131
50	1·6907	300	10·1442	550	18·5977	800	27·0512
60	2·0288	310	10·4823	560	18·9358	810	27·3893
70	2·3670	320	10·8205	570	19·2740	820	27·7275
80	2·7051	330	11·1586	580	19·6121	830	28·0656
90	3·0433	340	11·4968	590	19·9503	840	28·4038
100	3·3814	350	11·8349	600	20·2884	850	28·7419
110	3·7195	360	12·1730	610	20·6265	860	29·0800
120	4·0577	370	12·5112	620	20·9647	870	29·4182
130	4·3958	380	12·8493	630	21·3028	880	29·7563
140	4·7340	390	13·1875	640	21·6410	890	30·0945
150	5·0721	400	13·5256	650	21·9791	900	30·4326
160	5·4102	410	13·8637	660	22·3172	910	30·7707
170	5·7484	420	14·2019	670	22·6554	920	31·1089
180	6·0865	430	14·5400	680	22·9935	930	31·4470
190	6·4247	440	14·8782	690	23·3317	940	31·7852
200	6·7628	450	15·2163	700	23·6698	950	32·1233
210	7·1009	460	15·5544	710	24·0079	960	32·4614
220	7·4391	470	15·8926	720	24·3461	970	32·7996
230	7·7772	480	16·2307	730	24·6842	980	33·1377
240	8·1154	490	16·5689	740	25·0224	990	33·4759
250	8·4535	500	16·9070	750	25·3605	1000	33·8140

UK fluid ounces to decilitres

UK fl oz	dl	UK fl oz	dl	UK fl oz	dl	UK fl oz	dl
1	0·2841	26	7·3874	51	14·4907	76	21·5940
2	0·5683	27	7·6715	52	14·7748	77	21·8781
3	0·8524	28	7·9557	53	15·0589	78	22·1622
4	1·1365	29	8·2398	54	15·3431	79	22·4463
5	1·4207	30	8·5239	55	15·6272	80	22·7305
6	1·7048	31	8·8081	56	15·9113	81	23·0146
7	1·9889	32	9·0922	57	16·1955	82	23·2987
8	2·2730	33	9·3763	58	16·4796	83	23·5829
9	2·5572	34	9·6605	59	16·7637	84	23·8760
10	2·8413	35	9·9446	60	17·0479	85	24·1511
11	3·1254	36	10·2287	61	17·3320	86	24·4353
12	3·4096	37	10·5128	62	17·6161	87	24·7194
13	3·6937	38	10·7970	63	17·9003	88	25·0035
14	3·9778	39	11·0811	64	18·1844	89	25·2877
15	4·2620	40	11·3652	65	18·4685	90	25·5718
16	4·5461	41	11·6494	66	18·7526	91	25·8559
17	4·8302	42	11·9335	67	19·0368	92	26·1401
18	5·1144	43	12·2176	68	19·3209	93	26·4242
19	5·3985	44	12·5018	69	19·6050	94	26·7083
20	5·6826	45	12·7859	70	19·8892	95	26·9924
21	5·9668	46	13·0700	71	20·1733	96	27·2766
22	6·2509	47	13·3542	72	20·4574	97	27·5607
23	6·5350	48	13·6383	73	20·7416	98	27·8448
24	6·8191	49	13·9224	74	21·0257	99	28·1290
25	7·1033	50	14·2066	75	21·3098	100	28·4131

dl	UK fl oz	dl	UK fl oz	dl	UK fl oz	dl	UK fl oz
1	3·5195	26	91·5073	51	179·4950	76	267·4828
2	7·0390	27	95·0268	52	183·0145	77	271·0023
3	10·5585	28	98·5463	53	186·5340	78	274·5218
4	14·0780	29	102·0658	54	190·0535	79	278·0413
5	17·5976	30	105·5853	55	193·5731	80	281·5608
6	21·1171	31	109·1048	56	197·0926	81	285·0803
7	24·6366	32	112·6243	57	200·6121	82	288·5998
8	28·1561	33	116·1438	58	204·1316	83	292·1193
9	31·6756	34	119·6633	59	207·6511	84	295·6388
10	35·1951	35	123·1829	60	211·1706	85	299·1584
11	38·7146	36	126·7024	61	214·6901	86	302·6779
12	42·2341	37	130·2219	62	218·2096	87	306·1974
13	45·7536	38	133·7414	63	221·7291	88	309·7169
14	49·2731	39	137·2609	64	225·2486	89	313·2364
15	52·7927	40	140·7804	65	228·7682	90	316·7559
16	56·3122	41	144·2999	66	232·2877	91	320·2754
17	59·8317	42	147·8194	67	235·8072	92	323·7949
18	63·3512	43	151·3389	68	239·3267	93	327·3144
19	66·8707	44	154·8584	69	242·8462	94	330·8339
20	70·3902	45	158·3780	70	246·3657	95	334·3535
21	73·9097	46	161·8975	71	249·8852	96	337·8730
22	77·4292	47	165·4170	72	253·4047	97	341·3925
23	80·9487	48	168·9365	73	256·9242	98	344·9120
24	84·4682	49	172·4560	74	260·4437	99	348·4315
25	87·9878	50	175·9755	75	263·9633	100	351·9510

US fluid ounces
to decilitres

US fl oz	dl	US fl oz	dl	US fl oz	dl	US fl oz	dl
1	0·2957	26	7·6891	51	15·0825	76	22·4759
2	0·5915	27	7·9848	52	15·3782	77	22·7716
3	0·8872	28	8·2806	53	15·6740	78	23·0673
4	1·1829	29	8·5763	54	15·9697	79	23·3631
5	1·4787	30	8·8721	55	16·2654	80	23·6588
6	1·7744	31	9·1678	56	16·5612	81	23·9545
7	2·0701	32	9·4635	57	16·8569	82	24·2503
8	2·3659	33	9·7593	58	17·1526	83	24·5460
9	2·6616	34	10·0550	59	17·4484	84	24·8417
10	2·9574	35	10·3507	60	17·7441	85	25·1375
11	3·2531	36	10·6465	61	18·0398	86	25·4332
12	3·5488	37	10·9422	62	18·3356	87	25·7289
13	3·8446	38	11·2379	63	18·6313	88	26·0247
14	4·1403	39	11·5337	64	18·9270	89	26·3204
15	4·4360	40	11·8294	65	19·2228	90	26·6162
16	4·7318	41	12·1251	66	19·5185	91	26·9119
17	5·0275	42	12·4209	67	19·8142	92	27·2076
18	5·3232	43	12·7166	68	20·1100	93	27·5034
19	5·6190	44	13·0123	69	20·4057	94	27·7991
20	5·9147	45	13·3081	70	20·7015	95	28·0948
21	6·2104	46	13·6038	71	20·9972	96	28·3906
22	6·5062	47	13·8995	72	21·2929	97	28·6863
23	6·8019	48	14·1953	73	21·5887	98	28·9820
24	7·0976	49	14·4910	74	21·8844	99	29·2778
25	7·3934	50	14·7868	75	22·1801	100	29·5735

Decilitres
to US fluid ounces

dl	US fl oz	dl	US fl oz	dl	US fl oz	dl	US fl oz
1	3·3814	26	87·9164	51	172·4514	76	256·9864
2	6·7628	27	91·2978	52	175·8328	77	260·3678
3	10·1442	28	94·6792	53	179·2142	78	263·7492
4	13·5256	29	98·0606	54	182·5956	79	267·1306
5	16·9070	30	101·4420	55	185·9770	80	270·5120
6	20·2884	31	104·8234	56	189·3584	81	273·8934
7	23·6698	32	108·2048	57	192·7398	82	277·2748
8	27·0512	33	111·5862	58	196·1212	83	280·6562
9	30·4326	34	114·9676	59	199·5026	84	284·0376
10	33·8140	35	118·3490	60	202·8840	85	287·4190
11	37·1954	36	121·7304	61	206·2654	86	290·8004
12	40·5768	37	125·1118	62	209·6468	87	294·1818
13	43·9582	38	128·4932	63	213·0282	88	297·5632
14	47·3396	39	131·8746	64	216·4096	89	300·9446
15	50·7210	40	135·2560	65	219·7910	90	304·3260
16	54·1024	41	138·6374	66	223·1724	91	307·7074
17	57·4838	42	142·0188	67	226·5538	92	311·0888
18	60·8652	43	145·4002	68	229·9352	93	314·4702
19	64·2466	44	148·7816	69	233·3166	94	317·8516
20	67·6280	45	152·1630	70	236·6980	95	321·2330
21	71·0094	46	155·5444	71	240·0794	96	324·6144
22	74·3908	47	158·9258	72	243·4608	97	327·9958
23	77·7722	48	162·3072	73	246·8422	98	331·3772
24	81·1536	49	165·6886	74	250·2236	99	334·7586
25	84·5350	50	169·0700	75	253·6050	100	338·1400

UK pints
to litres

UK pt	l	UK pt	l	UK pt	l	UK pt	l
1	0·5683	26	14·7748	51	28·9813	76	43·1878
2	1·1365	27	15·3430	52	29·5496	77	43·7561
3	1·7048	28	15·9113	53	30·1178	78	44·3244
4	2·2730	29	16·4796	54	30·6861	79	44·8926
5	2·8413	30	17·0478	55	31·2544	80	45·4609
6	3·4096	31	17·6161	56	31·8226	81	46·0291
7	3·9778	32	18·1844	57	32·3909	82	46·5974
8	4·5461	33	18·7526	58	32·9591	83	47·1657
9	5·1143	34	19·3209	59	33·5274	84	47·7339
10	5·6826	35	19·8891	60	34·0957	85	48·3022
11	6·2509	36	20·4574	61	34·6639	86	48·8704
12	6·8191	37	21·0257	62	35·2322	87	49·4387
13	7·3874	38	21·5939	63	35·8004	88	50·0070
14	7·9557	39	22·1622	64	36·3687	89	50·5752
15	8·5239	40	22·7304	65	36·9370	90	51·1435
16	9·0922	41	23·2987	66	37·5052	91	51·7118
17	9·6604	42	23·8670	67	38·0735	92	52·2800
18	10·2287	43	24·4352	68	38·6417	93	52·8483
19	10·7970	44	25·0035	69	39·2100	94	53·4165
20	11·3652	45	25·5717	70	39·7783	95	53·9848
21	11·9335	46	26·1400	71	40·3465	96	54·5531
22	12·5017	47	26·7083	72	40·9148	97	55·1213
23	13·0700	48	27·2765	73	41·4831	98	55·6896
24	13·6383	49	27·8448	74	42·0513	99	56·2578
25	14·2065	50	28·4131	75	42·6196	100	56·8261

UK pt	l	UK pt	l	UK pt	l	UK pt	l
110	62·5087	360	204·5740	610	346·6392	860	488·7045
120	68·1913	370	210·2566	620	352·3218	870	494·3871
130	73·8739	380	215·9392	630	358·0044	880	500·0697
140	79·5565	390	221·6218	640	363·6870	890	505·7523
150	85·2392	400	227·3044	650	369·3697	900	511·4349
160	90·9218	410	232·9870	660	375·0523	910	517·1175
170	96·6044	420	238·6696	670	380·7349	920	522·8001
180	102·2870	430	244·3522	680	386·4175	930	528·4827
190	107·9696	440	250·0348	690	392·1001	940	534·1653
200	113·6522	450	255·7175	700	397·7827	950	539·8480
210	119·3348	460	261·4001	710	403·4653	960	545·5306
220	125·0174	470	267·0827	720	409·1479	970	551·2132
230	130·7000	480	272·7653	730	414·8305	980	556·8958
240	136·3826	490	278·4479	740	420·5131	990	562·5784
250	142·0653	500	284·1305	750	426·1958	1000	568·2610
260	147·7479	510	289·8131	760	431·8784		
270	153·4305	520	295·4957	770	437·5610		
280	159·1131	530	301·1783	780	443·2436		
290	164·7957	540	306·8609	790	448·9262		
300	170·4783	550	312·5436	800	454·6088		
310	176·1609	560	318·2262	810	460·2914		
320	181·8435	570	323·9088	820	465·9740		
330	187·5261	580	329·5914	830	471·6566		
340	193·2087	590	335·2740	840	477·3392		
350	198·8914	600	340·9566	850	483·0219		

Litres
to UK pints

l	UK pt	l	UK pt	l	UK pt	l	UK pt
1	1·7598	26	45·7535	51	89·7473	76	133·7410
2	3·5195	27	47·5133	52	91·5070	77	135·5008
3	5·2793	28	49·2730	53	93·2668	78	137·2605
4	7·0390	29	51·0328	54	95·0265	79	139·0203
5	8·7988	30	52·7925	55	96·7863	80	140·7800
6	10·5585	31	54·5523	56	98·5460	81	142·5398
7	12·3183	32	56·3120	57	100·3058	82	144·2995
8	14·0780	33	58·0718	58	102·0655	83	146·0593
9	15·8378	34	59·8315	59	103·8253	84	147·8190
10	17·5975	35	61·5913	60	105·5850	85	149·5788
11	19·3573	36	63·3510	61	107·3448	86	151·3385
12	21·1170	37	65·1108	62	109·1045	87	153·0983
13	22·8768	38	66·8705	63	110·8643	88	154·8580
14	24·6365	39	68·6303	64	112·6240	89	156·6178
15	26·3963	40	70·3900	65	114·3838	90	158·3775
16	28·1560	41	72·1498	66	116·1435	91	160·1373
17	29·9158	42	73·9095	67	117·9033	92	161·8970
18	31·6755	43	75·6693	68	119·6630	93	163·6568
19	33·4353	44	77·4290	69	121·4228	94	165·4165
20	35·1950	45	79·1888	70	123·1825	95	167·1763
21	36·9548	46	80·9485	71	124·9423	96	168·9360
22	38·7145	47	82·7083	72	126·7020	97	170·6958
23	40·4743	48	84·4680	73	128·4618	98	172·4555
24	42·2340	49	86·2278	74	130·2215	99	174·2153
25	43·9938	50	87·9875	75	131·9813	100	175·9750

Litres
to UK pints

l	UK pt	l	UK pt	l	UK pt	l	UK pt
110	193·5725	360	633·5100	610	1073·4475	860	1513·3850
120	211·1700	370	651·1075	620	1091·0450	870	1530·9825
130	228·7675	380	668·7050	630	1108·6425	880	1548·5800
140	246·3650	390	686·3025	640	1126·2400	890	1566·1775
150	263·9625	400	703·9000	650	1143·8375	900	1583·7750
160	281·5600	410	721·4975	660	1161·4350	910	1601·3725
170	299·1575	420	739·0950	670	1179·0325	920	1618·9700
180	316·7550	430	756·6925	680	1196·6300	930	1636·5675
190	334·3525	440	774·2900	690	1214·2275	940	1654·1650
200	351·9500	450	791·8875	700	1231·8250	950	1671·7625
210	369·5475	460	809·4850	710	1249·4225	960	1689·3600
220	387·1450	470	827·0825	720	1267·0200	970	1706·9575
230	404·7425	480	844·6800	730	1284·6175	980	1724·5550
240	422·3400	490	862·2775	740	1302·2150	990	1742·1525
250	439·9375	500	879·8750	750	1319·8125	1000	1759·7500
260	457·5350	510	897·4725	760	1337·4100		
270	475·1325	520	915·0700	770	1355·0075		
280	492·7300	530	932·6675	780	1372·6050		
290	510·3275	540	950·2650	790	1390·2025		
300	527·9250	550	967·8625	800	1407·8000		
310	545·5225	560	985·4600	810	1425·3975		
320	563·1200	570	1003·0575	820	1442·9950		
330	580·7175	580	1020·6550	830	1460·5925		
340	598·3150	590	1038·2525	840	1478·1900		
350	615·9125	600	1055·8500	850	1495·7875		

US liquid pints
to litres

US liq pt	l	US liq pt	l	US liq pt	l	US liq pt	l
1	0·4732	26	12·3026	51	24·1320	76	35·9614
2	0·9464	27	12·7758	52	24·6052	77	36·4346
3	1·4195	28	13·2489	53	25·0783	78	36·9077
4	1·8927	29	13·7221	54	25·5515	79	37·3809
5	2·3659	30	14·1953	55	26·0247	80	37·8541
6	2·8391	31	14·6685	56	26·4979	81	38·3273
7	3·3122	32	15·1416	57	26·9710	82	38·8004
8	3·7854	33	15·6148	58	27·4442	83	39·2736
9	4·2586	34	16·0880	59	27·9174	84	39·7468
10	4·7318	35	16·5612	60	28·3906	85	40·2200
11	5·2049	36	17·0343	61	28·8637	86	40·6931
12	5·6781	37	17·5075	62	29·3369	87	41·1663
13	6·1513	38	17·9807	63	29·8101	88	41·6395
14	6·6245	39	18·4539	64	30·2833	89	42·1127
15	7·0976	40	18·9270	65	30·7564	90	42·5858
16	7·5708	41	19·4002	66	31·2296	91	43·0590
17	8·0440	42	19·8734	67	31·7028	92	43·5322
18	8·5172	43	20·3466	68	32·1760	93	44·0054
19	8·9903	44	20·8197	69	32·6491	94	44·4785
20	9·4635	45	21·2929	70	33·1223	95	44·9517
21	9·9367	46	21·7661	71	33·5955	96	45·4249
22	10·4099	47	22·2393	72	34·0687	97	45·8981
23	10·8830	48	22·7124	73	34·5418	98	46·3712
24	11·3562	49	23·1856	74	35·0150	99	46·8444
25	11·8294	50	23·6588	75	35·4882	100	47·3176

US liquid pints to litres

US liq pt	l	US liq pt	l	US liq pt	l	US liq pt	l
110	52·0494	360	170·3434	610	288·6374	860	406·9314
120	56·7811	370	175·0751	620	293·3691	870	411·6631
130	61·5129	380	179·8069	630	298·1009	880	416·3949
140	66·2446	390	184·5386	640	302·8326	890	421·1266
150	70·9764	400	189·2704	650	307·5644	900	425·8584
160	75·7082	410	194·0022	660	312·2962	910	430·5902
170	80·4399	420	198·7339	670	317·0279	920	435·3219
180	85·1717	430	203·4657	680	321·7597	930	440·0537
190	89·9034	440	208·1974	690	326·4914	940	444·7854
200	94·6352	450	212·9292	700	331·2232	950	449·5172
210	99·3670	460	217·6610	710	335·9550	960	454·2490
220	104·0987	470	222·3927	720	340·6867	970	458·9807
230	108·8305	480	227·1245	730	345·4185	980	463·7125
240	113·5622	490	231·8562	740	350·1502	990	468·4442
250	118·2940	500	236·5880	750	354·8820	1000	473·1760
260	123·0258	510	241·3198	760	359·6138		
270	127·7575	520	246·0515	770	364·3455		
280	132·4893	530	250·7833	780	369·0773		
290	137·2210	540	255·5150	790	373·8090		
300	141·9528	550	260·2468	800	378·5408		
310	146·6846	560	264·9786	810	383·2726		
320	151·4163	570	269·7103	820	388·0043		
330	156·1481	580	274·4421	830	392·7361		
340	160·8798	590	279·1738	840	397·4678		
350	165·6116	600	283·9056	850	402·1996		

Litres
to US liquid pints

l	US liq pt	l	US liq pt	l	US liq pt	l	US liq pt
1	2·1134	26	54·9479	51	107·7824	76	160·6169
2	4·2268	27	57·0613	52	109·8958	77	162·7303
3	6·3401	28	59·1746	53	112·0091	78	164·8436
4	8·4535	29	61·2880	54	114·1225	79	166·9570
5	10·5669	30	63·4014	55	116·2359	80	169·0704
6	12·6803	31	65·5148	56	118·3493	81	171·1838
7	14·7937	32	67·6282	57	120·4627	82	173·2972
8	16·9070	33	69·7415	58	122·5760	83	175·4105
9	19·0204	34	71·8549	59	124·6894	84	177·5239
10	21·1338	35	73·9683	60	126·8028	85	179·6373
11	23·2472	36	76·0817	61	128·9162	86	181·7507
12	25·3606	37	78·1951	62	131·0296	87	183·8641
13	27·4739	38	80·3084	63	133·1429	88	185·9774
14	29·5873	39	82·4218	64	135·2563	89	188·0908
15	31·7007	40	84·5352	65	137·3697	90	190·2042
16	33·8141	41	86·6486	66	139·4831	91	192·3176
17	35·9275	42	88·7620	67	141·5965	92	194·4310
18	38·0408	43	90·8753	68	143·7098	93	196·5443
19	40·1542	44	92·9887	69	145·8232	94	198·6577
20	42·2676	45	95·1021	70	147·9366	95	200·7711
21	44·3810	46	97·2155	71	150·0500	96	202·8845
22	46·4944	47	99·3289	72	152·1634	97	204·9979
23	48·6077	48	101·4422	73	154·2767	98	207·1112
24	50·7211	49	103·5556	74	156·3901	99	209·2246
25	52·8345	50	105·6690	75	158·5035	100	211·3380

Litres
to US liquid pints

l	US liq pt	l	US liq pt	l	US liq pt	l	US liq pt
110	232·4718	360	760·8168	610	1289·1618	860	1817·5068
120	253·6056	370	781·9506	620	1310·2956	870	1838·6406
130	274·7394	380	803·0844	630	1331·4294	880	1859·7744
140	295·8732	390	824·2182	640	1352·5632	890	1880·9082
150	317·0070	400	845·3520	650	1373·6970	900	1902·0420
160	338·1408	410	866·4858	660	1394·8308	910	1923·1758
170	359·2746	420	887·6196	670	1415·9646	920	1944·3096
180	380·4084	430	908·7534	680	1437·0984	930	1965·4434
190	401·5422	440	929·8872	690	1458·2322	940	1986·5772
200	422·6760	450	951·0210	700	1479·3660	950	2007·7110
210	443·8098	460	972·1548	710	1500·4998	960	2028·8448
220	464·9436	470	993·2886	720	1521·6336	970	2049·9786
230	486·0774	480	1014·4224	730	1542·7674	980	2071·1124
240	507·2112	490	1035·5562	740	1563·9012	990	2092·2462
250	528·3450	500	1056·6900	750	1585·0350	1000	2113·3800
260	549·4788	510	1077·8238	760	1606·1688		
270	570·6126	520	1098·9576	770	1627·3026		
280	591·7464	530	1120·0914	780	1648·4364		
290	612·8802	540	1141·2252	790	1669·5702		
300	634·0140	550	1162·3590	800	1690·7040		
310	655·1478	560	1183·4928	810	1711·8378		
320	676·2816	570	1204·6266	820	1732·9716		
330	697·4154	580	1225·7604	830	1754·1054		
340	718·5492	590	1246·8942	840	1775·2392		
350	739·6830	600	1268·0280	850	1796·3730		

UK gallons
to litres

UK gal	l	UK gal	l	UK gal	l	UK gal	l
1	4·546	26	118·198	51	231·851	76	345·503
2	9·092	27	122·744	52	236·397	77	350·049
3	13·638	28	127·291	53	240·943	78	354·595
4	18·184	29	131·837	54	245·489	79	359·141
5	22·731	30	136·383	55	250·035	80	363·687
6	27·277	31	140·929	56	254·581	81	368·233
7	31·823	32	145·475	57	259·127	82	372·779
8	36·369	33	150·021	58	263·673	83	377·325
9	40·915	34	154·567	59	268·219	84	381·872
10	45·461	35	159·113	60	272·765	85	386·418
11	50·007	36	163·659	61	277·311	86	390·964
12	54·553	37	168·205	62	281·858	87	395·510
13	59·099	38	172·751	63	286·404	88	400·056
14	63·645	39	177·298	64	290·950	89	404·602
15	68·191	40	181·844	65	295·496	90	409·148
16	72·737	41	186·390	66	300·042	91	413·694
17	77·284	42	190·936	67	304·588	92	418·240
18	81·830	43	195·482	68	309·134	93	422·786
19	86·376	44	200·028	69	313·680	94	427·332
20	90·922	45	204·574	70	318·226	95	431·879
21	95·468	46	209·120	71	322·772	96	436·425
22	100·014	47	213·666	72	327·318	97	440·971
23	104·560	48	218·212	73	331·865	98	445·517
24	109·106	49	222·758	74	336·411	99	450·063
25	113·652	50	227·305	75	340·957	100	454·609

UK gallons to litres

UK gal	l	UK gal	l	UK gal	l	UK gal	l
110	500·070	360	1636·594	610	2773·118	860	3909·642
120	545·531	370	1682·055	620	2818·579	870	3955·103
130	590·992	380	1727·516	630	2864·040	880	4000·564
140	636·453	390	1772·977	640	2909·501	890	4046·025
150	681·914	400	1818·438	650	2954·962	900	4091·486
160	727·375	410	1863·899	660	3000·423	910	4136·946
170	772·836	420	1909·360	670	3045·884	920	4182·407
180	818·297	430	1954·821	680	3091·345	930	4227·868
190	863·758	440	2000·282	690	3136·806	940	4273·329
200	909·219	450	2045·743	700	3182·267	950	4318·790
210	954·680	460	2091·204	710	3227·727	960	4364·251
220	1000·141	470	2136·665	720	3273·188	970	4409·712
230	1045·602	480	2182·126	730	3318·649	980	4455·173
240	1091·063	490	2227·587	740	3364·110	990	4500·634
250	1136·524	500	2273·048	750	3409·571	1000	4546·095
260	1181·985	510	2318·508	760	3455·032	2000	9092·190
270	1227·446	520	2363·969	770	3500·493	3000	13638·285
280	1272·907	530	2409·430	780	3545·954	4000	18184·380
290	1318·368	540	2454·891	790	3591·415	5000	22730·475
300	1363·829	550	2500·352	800	3636·876	6000	27276·570
310	1409·289	560	2545·813	810	3682·337	7000	31822·665
320	1454·750	570	2591·274	820	3727·798	8000	36368·760
330	1500·211	580	2636·735	830	3773·259	9000	40914·855
340	1545·672	590	2682·196	840	3818·720	10000	45460·950
350	1591·133	600	2727·657	850	3864·181	20000	90921·900

Litres
to UK gallons

l	UK gal	l	UK gal	l	UK gal	l	UK gal
1	0·2200	26	5·7192	51	11·2184	76	16·7176
2	0·4399	27	5·9392	52	11·4384	77	16·9376
3	0·6599	28	6·1591	53	11·6584	78	17·1576
4	0·8799	29	6·3791	54	11·8783	79	17·3776
5	1·0998	30	6·5991	55	12·0983	80	17·5975
6	1·3198	31	6·8190	56	12·3183	81	17·8175
7	1·5398	32	7·0390	57	12·5382	82	18·0375
8	1·7598	33	7·2590	58	12·7582	83	18·2574
9	1·9797	34	7·4789	59	12·9782	84	18·4774
10	2·1997	35	7·6989	60	13·1981	85	18·6974
11	2·4197	36	7·9189	61	13·4181	86	18·9173
12	2·6396	37	8·1389	62	13·6381	87	19·1373
13	2·8596	38	8·3588	63	13·8580	88	19·3573
14	3·0796	39	8·5788	64	14·0780	89	19·5772
15	3·2995	40	8·7988	65	14·2980	90	19·7972
16	3·5195	41	9·0187	66	14·5180	91	20·0172
17	3·7395	42	9·2387	67	14·7379	92	20·2371
18	3·9595	43	9·4587	68	14·9579	93	20·4571
19	4·1794	44	9·6786	69	15·1779	94	20·6771
20	4·3994	45	9·8986	70	15·3978	95	20·8971
21	4·6193	46	10·1186	71	15·6178	96	21·1170
22	4·8393	47	10·3385	72	15·8378	97	21·3370
23	5·0593	48	10·5585	73	16·0578	98	21·5570
24	5·2793	49	10·7785	74	16·2777	99	21·7769
25	5·4992	50	10·9985	75	16·4977	100	21·9969

Litres
to UK gallons

l	UK gal	l	UK gal	l	UK gal	l	UK gal
110	24·1966	360	79·1888	610	134·1811	860	189·1733
120	26·3963	370	81·3885	620	136·3808	870	191·3730
130	28·5960	380	83·5882	630	138·5805	880	193·5727
140	30·7957	390	85·7879	640	140·7802	890	195·7724
150	32·9954	400	87·9876	650	142·9799	900	197·9721
160	35·1950	410	90·1873	660	145·1795	910	200·1718
170	37·3947	420	92·3870	670	147·3792	920	202·3715
180	39·5944	430	94·5867	680	149·5789	930	204·5712
190	41·7941	440	96·7864	690	151·7786	940	206·7709
200	43·9938	450	98·9861	700	153·9783	950	208·9706
210	46·1935	460	101·1857	710	156·1780	960	211·1702
220	48·3932	470	103·3854	720	158·3777	970	213·3699
230	50·5929	480	105·5851	730	160·5774	980	215·5696
240	52·7926	490	107·7848	740	162·7771	990	217·7693
250	54·9923	500	109·9845	750	164·9768	1000	219·9690
260	57·1919	510	112·1842	760	167·1764	2000	439·9380
270	59·3916	520	114·3839	770	169·3761	3000	659·9070
280	61·5913	530	116·5836	780	171·5758	4000	879·8760
290	63·7910	540	118·7833	790	173·7755	5000	1099·8450
300	65·9907	550	120·9830	800	175·9752	6000	1319·8140
310	68·1904	560	123·1826	810	178·1749	7000	1539·7830
320	70·3901	570	125·3823	820	180·3746	8000	1759·7520
330	72·5898	580	127·5820	830	182·5743	9000	1979·7210
340	74·7895	590	129·7817	840	184·7740	10000	2199·6900
350	76·9892	600	131·9814	850	186·9737	20000	4399·3800

US gallons
to litres

US gal	l	US gal	l	US gal	l	US gal	l
1	3·785	26	98·421	51	193·056	76	287·691
2	7·571	27	102·206	52	196·841	77	291·477
3	11·356	28	105·992	53	200·627	78	295·262
4	15·142	29	109·777	54	204·412	79	299·047
5	18·927	30	113·562	55	208·198	80	302·832
6	22·713	31	117·348	56	211·983	81	306·618
7	26·498	32	121·133	57	215·768	82	310·404
8	30·283	33	124·919	58	219·554	83	314·189
9	34·069	34	128·704	59	223·339	84	317·974
10	37·854	35	132·489	60	227·125	85	321·760
11	41·640	36	136·275	61	230·910	86	325·545
12	45·425	37	140·060	62	234·695	87	329·331
13	49·210	38	143·846	63	238·481	88	333·116
14	52·996	39	147·631	64	242·266	89	336·902
15	56·781	40	151·416	65	246·052	90	340·687
16	60·567	41	155·202	66	249·837	91	344·472
17	64·352	42	158·987	67	253·623	92	348·258
18	68·137	43	162·773	68	257·408	93	352·043
19	71·923	44	166·558	69	261·193	94	355·829
20	75·708	45	170·346	70	264·979	95	359·614
21	79·494	46	174·129	71	268·764	96	363·399
22	83·279	47	177·914	72	272·550	97	367·185
23	87·064	48	181·700	73	276·335	98	370·970
24	90·850	49	185·485	74	280·120	99	374·756
25	94·635	50	189·271	75	283·906	100	378·541

US gallons to litres

US gal	l	US gal	l	US gal	l	US gal	l
110	416·395	360	1362·748	610	2309·100	860	3255·453
120	454·249	370	1400·602	620	2346·954	870	3293·307
130	492·103	380	1438·456	630	2384·808	880	3331·161
140	529·957	390	1476·310	640	2422·662	890	3369·015
150	567·812	400	1514·164	650	2460·517	900	3406·869
160	605·666	410	1552·018	660	2498·371	910	3444·723
170	643·520	420	1589·872	670	2536·225	920	3482·577
180	681·374	430	1627·726	680	2574·079	930	3520·431
190	719·228	440	1665·580	690	2611·933	940	3557·285
200	757·082	450	1703·435	700	2649·787	950	3596·140
210	794·936	460	1741·289	710	2687·641	960	3633·994
220	832·790	470	1779·143	720	2725·495	970	3671·848
230	870·644	480	1816·997	730	2763·349	980	3709·702
240	908·498	490	1854·851	740	2801·203	990	3747·556
250	946·353	500	1892·705	750	2839·058	1000	3785·410
260	984·207	510	1930·559	760	2876·912	2000	7570·820
270	1022·061	520	1968·413	770	2914·766	3000	11356·230
280	1059·915	530	2006·267	780	2952·620	4000	15141·640
290	1097·769	540	2044·121	790	2990·474	5000	18927·050
300	1135·623	550	2081·976	800	3028·328	6000	22712·460
310	1173·477	560	2119·830	810	3066·182	7000	26497·870
320	1211·331	570	2157·684	820	3104·036	8000	30283·280
330	1249·185	580	2195·538	830	3141·890	9000	34068·690
340	1287·039	590	2233·392	840	3179·744	10000	37854·100
350	1324·894	600	2271·246	850	3217·599	25000	94635·250

Litres
to US gallons

l	US gal	l	US gal	l	US gal	l	US gal
1	0·2642	26	6·8685	51	13·4728	76	20·0771
2	0·5283	27	7·1326	52	13·7369	77	20·3412
3	0·7925	28	7·3968	53	14·0011	78	20·6054
4	1·0567	29	7·6610	54	14·2653	79	20·8696
5	1·3209	30	7·9252	55	14·5295	80	21·1338
6	1·5850	31	8·1893	56	14·7936	81	21·3979
7	1·8492	32	8·4535	57	15·0578	82	21·6621
8	2·1134	33	8·7177	58	15·3220	83	21·9263
9	2·3775	34	8·9818	59	15·5861	84	22·1904
10	2·6417	35	9·2460	60	15·8503	85	22·4546
11	2·9059	36	9·5102	61	16·1145	86	22·7188
12	3·1701	37	9·7744	62	16·3787	87	22·9830
13	3·4342	38	10·0385	63	16·6428	88	23·2471
14	3·6984	39	10·3027	64	16·9070	89	23·5113
15	3·9626	40	10·5669	65	17·1712	90	23·7755
16	4·2268	41	10·8311	66	17·4354	91	24·0397
17	4·4909	42	11·0952	67	17·6995	92	24·3038
18	4·7551	43	11·3594	68	17·9637	93	24·5680
19	5·0193	44	11·6236	69	18·2279	94	24·8322
20	5·2834	45	11·8877	70	18·4920	95	25·0963
21	5·5476	46	12·1519	71	18·7562	96	25·3605
22	5·8118	47	12·4161	72	19·0204	97	25·6247
23	6·0760	48	12·6803	73	19·2846	98	25·8889
24	6·3401	49	12·9444	74	19·5487	99	26·1530
25	6·6043	50	13·2086	75	19·8129	100	26·4172

Litres
to US gallons

l	US gal	l	US gal	l	US gal	l	US gal
110	29·0589	360	95·1019	610	161·1449	860	227·1879
120	31·7006	370	97·7436	620	163·7866	870	229·8296
130	34·3424	380	100·3854	630	166·4284	880	232·4714
140	36·9841	390	103·0271	640	169·0701	890	235·1131
150	39·6258	400	105·6688	650	171·7118	900	237·7548
160	42·2675	410	108·3105	660	174·3535	910	240·3965
170	44·9092	420	110·9522	670	176·9952	920	243·0382
180	47·5510	430	113·5940	680	179·6370	930	245·6800
190	50·1927	440	116·2357	690	182·2787	940	248·3217
200	52·8344	450	118·8774	700	184·9204	950	250·9634
210	55·4761	460	121·5191	710	187·5621	960	253·6051
220	58·1178	470	124·1608	720	190·2038	970	256·2468
230	60·7596	480	126·8026	730	192·8456	980	258·8886
240	63·4013	490	129·4443	740	195·4873	990	261·5303
250	66·0430	500	132·0860	750	198·1290	1000	264·1720
260	68·6847	510	134·7277	760	200·7707	2000	528·3440
270	71·3264	520	137·3694	770	203·4124	3000	792·5160
280	73·9682	530	140·0112	780	206·0542	4000	1056·6880
290	76·6099	540	142·6529	790	208·6959	5000	1320·8600
300	79·2516	550	145·2946	800	211·3376	6000	1585·0320
310	81·8933	560	147·9363	810	213·9793	7000	1849·2040
320	84·5350	570	150·5780	820	216·6210	8000	2113·3760
330	87·1768	580	153·2198	830	219·2628	9000	2377·5480
340	89·8185	590	155·8615	840	221·9045	10000	2641·7200
350	92·4602	600	158·5032	850	224·5462	25000	6604·3000

UK gallons
to US gallons

UK gal	US gal	UK gal	US gal	UK gal	US gal	UK gal	US gal
1	1·2010	26	31·2247	51	61·2485	76	91·2722
2	2·4019	27	32·4257	52	62·4494	77	92·4732
3	3·6029	28	33·6266	53	63·6504	78	93·6741
4	4·8038	29	34·8276	54	64·8513	79	94·8751
5	6·0048	30	36·0285	55	66·0523	80	96·0760
6	7·2057	31	37·2295	56	67·2532	81	97·2770
7	8·4067	32	38·4304	57	68·4542	82	98·4779
8	9·6076	33	39·6314	58	69·6551	83	99·6789
9	10·8086	34	40·8323	59	70·8561	84	100·8798
10	12·0095	35	42·0333	60	72·0570	85	102·0808
11	13·2105	36	43·2342	61	73·2580	86	103·2817
12	14·4114	37	44·4352	62	74·4589	87	104·4827
13	15·6124	38	45·6361	63	75·6599	88	105·6836
14	16·8133	39	46·8371	64	76·8608	89	106·8846
15	18·0143	40	48·0380	65	78·0618	90	108·0855
16	19·2152	41	49·2390	66	79·2627	91	109·2865
17	20·4162	42	50·4399	67	80·4637	92	110·4874
18	21·6171	43	51·6409	68	81·6646	93	111·6884
19	22·8181	44	52·8418	69	82·8656	94	112·8893
20	24·0190	45	54·0428	70	84·0665	95	114·0903
21	25·2200	46	55·2437	71	85·2675	96	115·2912
22	26·4209	47	56·4447	72	86·4684	97	116·4922
23	27·6219	48	57·6456	73	87·6694	98	117·6931
24	28·8228	49	58·8466	74	88·8703	99	118·8941
25	30·0238	50	60·0475	75	90·0713	100	120·0950

UK gallons to US gallons

UK gal	US gal	UK gal	US gal	UK gal	US gal	UK gal	US gal
110	132·105	360	432·342	610	732·580	860	1032·817
120	144·114	370	444·352	620	744·589	870	1044·827
130	156·124	380	456·361	630	756·599	880	1056·836
140	168·133	390	468·371	640	768·608	890	1068·846
150	180·143	400	480·380	650	780·618	900	1080·855
160	192·152	410	492·390	660	792·627	910	1092·865
170	204·162	420	504·399	670	804·637	920	1104·874
180	216·171	430	516·409	680	816·646	930	1116·884
190	228·181	440	528·418	690	828·656	940	1128·893
200	240·190	450	540·428	700	840·665	950	1140·903
210	252·200	460	552·437	710	852·675	960	1152·912
220	264·209	470	564·447	720	864·684	970	1164·922
230	276·219	480	576·456	730	876·694	980	1176·931
240	288·228	490	588·466	740	888·703	990	1188·941
250	300·238	500	600·475	750	900·713	1000	1200·950
260	312·247	510	612·485	760	912·722	2000	2401·900
270	324·257	520	624·494	770	924·732	3000	3602·850
280	336·266	530	636·504	780	936·741	4000	4803·800
290	348·276	540	648·513	790	948·751	5000	6004·750
300	360·285	550	660·523	800	960·760	6000	7205·700
310	372·295	560	672·532	810	972·770	7000	8406·650
320	384·304	570	684·542	820	984·779	8000	9607·600
330	396·314	580	696·551	830	996·789	9000	10808·550
340	408·323	590	708·561	840	1008·798	10000	12009·500
350	420·333	600	720·570	850	1020·808	25000	30023·750

US gallons to UK gallons

US gal	UK gal	US gal	UK gal	US gal	UK gal	US gal	UK gal
1	0·8327	26	21·6495	51	42·4664	76	63·2832
2	1·6650	27	22·4822	52	43·2990	77	64·1159
3	2·4980	28	23·3149	53	44·1317	78	64·9486
4	3·3307	29	24·1475	54	44·9644	79	65·7812
5	4·1634	30	24·9802	55	45·7971	80	66·6139
6	4·9960	31	25·8129	56	46·6297	81	67·4466
7	5·8287	32	26·6456	57	47·4624	82	68·2793
8	6·6614	33	27·4782	58	48·2951	83	69·1119
9	7·4941	34	28·3109	59	49·1278	84	69·9446
10	8·3267	35	29·1436	60	49·9604	85	70·7773
11	9·1594	36	29·9763	61	50·7931	86	71·6100
12	9·9921	37	30·8089	62	51·6258	87	72·4426
13	10·8248	38	31·6416	63	52·4585	88	73·2753
14	11·6574	39	32·4743	64	53·2911	89	74·1080
15	12·4901	40	33·3070	65	54·1238	90	74·9407
16	13·3228	41	34·1396	66	54·9565	91	75·7733
17	14·1555	42	34·9723	67	55·7892	92	76·6060
18	14·9881	43	35·8050	68	56·6218	93	77·4387
19	15·8208	44	36·6377	69	57·4545	94	78·2714
20	16·6535	45	37·4703	70	58·2872	95	79·1040
21	17·4862	46	38·3030	71	59·1199	96	79·9367
22	18·3188	47	39·1357	72	59·9525	97	80·7694
23	19·1515	48	39·9684	73	60·7852	98	81·6021
24	19·9842	49	40·8010	74	61·6179	99	82·4347
25	20·8169	50	41·6337	75	62·4506	100	83·2674

US gallons to UK gallons

US gal	UK gal	US gal	UK gal	US gal	UK gal	US gal	UK gal
110	91·594	360	299·763	610	507·931	860	716·100
120	99·921	370	308·089	620	516·258	870	724·426
130	108·248	380	316·416	630	524·585	880	732·753
140	116·574	390	324·743	640	532·911	890	741·080
150	124·901	400	333·070	650	541·238	900	749·407
160	133·228	410	341·396	660	549·565	910	757·733
170	141·555	420	349·723	670	557·892	920	766·060
180	149·881	430	358·050	680	566·218	930	774·387
190	158·208	440	366·377	690	574·545	940	782·714
200	166·535	450	374·703	700	582·872	950	791·040
210	174·862	460	383·030	710	591·199	960	799·367
220	183·188	470	391·357	720	599·525	970	807·694
230	191·515	480	399·684	730	607·852	980	816·021
240	199·842	490	408·010	740	616·179	990	824·347
250	208·169	500	416·337	750	624·506	1000	832·674
260	216·495	510	424·664	760	632·832	2000	1665·348
270	224·822	520	432·990	770	641·159	3000	2498·022
280	233·149	530	441·317	780	649·486	4000	3330·696
290	241·475	540	449·644	790	657·812	5000	4163·370
300	249·802	550	457·971	800	666·139	6000	4996·044
310	258·129	560	466·297	810	674·466	7000	5828·718
320	266·456	570	474·624	820	682·793	8000	6661·392
330	274·782	580	482·951	830	691·119	9000	7494·066
340	283·109	590	491·278	840	699·446	10000	8326·740
350	291·436	600	499·604	850	707·773	25000	20816·850

US bushels
to hectolitres

US bu	hl	US bu	hl	US bu	hl	US bu	hl
1	0·3524	26	9·1622	51	17·9719	76	26·7817
2	0·7048	27	9·5146	52	18·3243	77	27·1341
3	1·0572	28	9·8669	53	18·6767	78	27·4865
4	1·4096	29	10·2193	54	19·0291	79	27·8389
5	1·7620	30	10·5717	55	19·3815	80	28·1913
6	2·1143	31	10·9241	56	19·7339	81	28·5437
7	2·4667	32	11·2765	57	20·0863	82	28·8961
8	2·8191	33	11·6289	58	20·4387	83	29·2485
9	3·1715	34	11·9813	59	20·7911	84	29·6008
10	3·5239	35	12·3337	60	21·1435	85	29·9532
11	3·8763	36	12·6861	61	21·4959	86	30·3056
12	4·2287	37	13·0385	62	21·8482	87	30·6580
13	4·5811	38	13·3909	63	22·2006	88	31·0104
14	4·9335	39	13·7432	64	22·5530	89	31·3628
15	5·2859	40	14·0956	65	22·9054	90	31·7152
16	5·6383	41	14·4480	66	23·2578	91	32·0676
17	5·9906	42	14·8004	67	23·6102	92	32·4200
18	6·3430	43	15·1528	68	23·9626	93	32·7724
19	6·6954	44	15·5052	69	24·3150	94	33·1248
20	7·0478	45	15·8576	70	24·6674	95	33·4771
21	7·4002	46	16·2100	71	25·0198	96	33·8295
22	7·7526	47	16·5624	72	25·3722	97	34·1819
23	8·1050	48	16·9148	73	25·7245	98	34·5343
24	8·4574	49	17·2672	74	26·0769	99	34·8867
25	8·8098	50	17·6196	75	26·4293	100	35·2391

US bu	hl	US bu	hl	US bu	hl	US bu	hl
110	38·7630	360	126·8608	610	214·9585	860	303·0563
120	42·2869	370	130·3847	620	218·4824	870	306·5802
130	45·8108	380	133·9086	630	222·0063	880	310·1041
140	49·3347	390	137·4325	640	225·5302	890	313·6280
150	52·8587	400	140·9564	650	229·0542	900	317·1519
160	56·3826	410	144·4803	660	232·5781	910	320·6758
170	59·9065	420	148·0042	670	236·1020	920	324·1997
180	63·4304	430	151·5281	680	239·6259	930	327·7236
190	66·9543	440	155·0520	690	243·1498	940	331·2475
200	70·4782	450	158·5760	700	246·6737	950	334·7715
210	74·0021	460	162·0999	710	250·1976	960	338·2954
220	77·5260	470	165·6238	720	253·7215	970	341·8193
230	81·0499	480	169·1477	730	257·2454	980	345·3432
240	84·5738	490	172·6716	740	260·7693	990	348·8671
250	88·0978	500	176·1955	750	264·2933	1000	352·3910
260	91·6217	510	179·7194	760	267·8172	2000	704·7820
270	95·1456	520	183·2433	770	271·3411	3000	1057·1730
280	98·6695	530	186·7672	780	274·8650	4000	1409·5640
290	102·1934	540	190·2911	790	278·3889	5000	1761·9550
300	105·7173	550	193·8151	800	281·9128	6000	2114·3460
310	109·2412	560	197·3390	810	285·4367	7000	2466·7370
320	112·7651	570	200·8629	820	288·9606	8000	2819·1280
330	116·2890	580	204·3868	830	292·4845	9000	3171·5190
340	119·8129	590	207·9107	840	296·0084	10000	3523·9100
350	123·3369	600	211·4346	850	299·5324	25000	8809·7750

Hectolitres
to US bushels

hl	US bu	hl	US bu	hl	US bu	hl	US bu
1	2·8378	26	73·7818	51	144·7258	76	215·6698
2	5·6755	27	76·6195	52	147·5635	77	218·5075
3	8·5133	28	79·4573	53	150·4013	78	221·3453
4	11·3510	29	82·2950	54	153·2390	79	224·1830
5	14·1888	30	85·1328	55	156·0768	80	227·0208
6	17·0266	31	87·9706	56	158·9146	81	229·8586
7	19·8643	32	90·8083	57	161·7523	82	232·6963
8	22·7021	33	93·6461	58	164·5901	83	235·5341
9	25·5398	34	96·4838	59	167·4278	84	238·3718
10	28·3776	35	99·3216	60	170·2656	85	241·2096
11	31·2154	36	102·1594	61	173·1034	86	244·0474
12	34·0531	37	104·9971	62	175·9411	87	246·8851
13	36·8909	38	107·8349	63	178·7789	88	249·7229
14	39·7286	39	110·6726	64	181·6166	89	252·5606
15	42·5664	40	113·5104	65	184·4544	90	255·3984
16	45·4042	41	116·3482	66	187·2922	91	258·2362
17	48·2419	42	119·1859	67	190·1299	92	261·0739
18	51·0797	43	122·0237	68	192·9677	93	263·9117
19	53·9174	44	124·8614	69	195·8054	94	266·7494
20	56·7552	45	127·6992	70	198·6432	95	269·5872
21	59·5930	46	130·5370	71	201·4810	96	272·4250
22	62·4307	47	133·3747	72	204·3187	97	275·2627
23	65·2685	48	136·2125	73	207·1565	98	278·1005
24	68·1062	49	139·0502	74	209·9942	99	280·9382
25	70·9440	50	141·8880	75	212·8320	100	283·7760

hl	US bu	hl	US bu	hl	US bu	hl	US bu
110	312·154	360	1021·594	610	1731·034	860	2440·474
120	340·531	370	1049·971	620	1759·411	870	2468·851
130	368·909	380	1078·349	630	1787·789	880	2497·229
140	397·286	390	1106·726	640	1816·166	890	2525·606
150	425·664	400	1135·104	650	1844·544	900	2553·984
160	454·042	410	1163·482	660	1872·922	910	2582·362
170	482·419	420	1191·859	670	1901·299	920	2610·739
180	510·797	430	1220·237	680	1929·677	930	2639·117
190	539·174	440	1248·614	690	1958·054	940	2667·494
200	567·552	450	1276·992	700	1986·432	950	2695·872
210	595·930	460	1305·370	710	2014·810	960	2724·250
220	624·307	470	1333·747	720	2043·187	970	2752·627
230	652·685	480	1362·125	730	2071·565	980	2781·005
240	681·062	490	1390·502	740	2099·942	990	2809·382
250	709·440	500	1418·880	750	2128·320	1000	2837·760
260	737·818	510	1447·258	760	2156·698	2000	5675·520
270	766·195	520	1475·635	770	2185·075	3000	8513·280
280	794·573	530	1504·013	780	2213·453	4000	11351·040
290	822·950	540	1532·390	790	2241·830	5000	14188·800
300	851·328	550	1560·768	800	2270·208	6000	17026·560
310	879·706	560	1589·146	810	2298·586	7000	19864·320
320	908·083	570	1617·523	820	2326·963	8000	22702·080
330	936·461	580	1645·901	830	2355·341	9000	25539·840
340	964·838	590	1674·278	840	2383·718	10000	28377·600
350	993·216	600	1702·656	850	2412·096	25000	70944·000

Grains
to grams

gr	g	gr	g	gr	g	gr	g
10	0·6480	260	16·8477	510	33·0474	760	49·2472
20	1·2960	270	17·4957	520	33·6954	770	49·8952
30	1·9440	280	18·1437	530	34·3434	780	50·5431
40	2·5920	290	18·7917	540	34·9914	790	51·1911
50	3·2400	300	19·4397	550	35·6394	800	51·8391
60	3·8879	310	20·0877	560	36·2874	810	52·4871
70	4·5359	320	20·7357	570	36·9354	820	53·1351
80	5·1839	330	21·3836	580	37·5834	830	53·7831
90	5·8319	340	22·0316	590	38·2314	840	54·4311
100	6·4799	350	22·6796	600	38·8793	850	55·0791
110	7·1279	360	23·3276	610	39·5273	860	55·7271
120	7·7759	370	23·9756	620	40·1753	870	56·3751
130	8·4237	380	24·6236	630	40·8233	880	57·0230
140	9·0719	390	25·2716	640	41·4713	890	57·6710
150	9·7198	400	25·9196	650	42·1193	900	58·3190
160	10·3678	410	26·5676	660	42·7673	910	58·9670
170	11·0158	420	27·2155	670	43·4153	920	59·6150
180	11·6638	430	27·8635	680	44·0633	930	60·2630
190	12·3118	440	28·5115	690	44·7112	940	60·9110
200	12·9598	450	29·1595	700	45·3592	950	61·5590
210	13·6078	460	29·8075	710	46·0072	960	62·2070
220	14·2558	470	30·4555	720	46·6552	970	62·8549
230	14·9037	480	31·1035	730	47·3032	980	63·5029
240	15·5517	490	31·7515	740	47·9512	990	64·1509
250	16·1997	500	32·3995	750	48·5992	1000	64·7989

gr	g	gr	g	gr	g	gr	g
1100	71·279	3600	233·276	6100	395·274	8600	557·271
1200	77·759	3700	239·756	6200	401·754	8700	563·751
1300	84·239	3800	246·236	6300	408·234	8800	570·231
1400	90·719	3900	252·716	6400	414·714	8900	576·711
1500	97·199	4000	259·196	6500	421·194	9000	583·191
1600	103·678	4100	265·676	6600	427·673	9100	589·671
1700	110·158	4200	272·156	6700	434·153	9200	596·151
1800	116·638	4300	278·636	6800	440·633	9300	602·631
1900	123·118	4400	285·116	6900	447·113	9400	609·111
2000	129·598	4500	291·596	7000	453·593	9500	615·591
2100	136·078	4600	298·075	7100	460·073	9600	622·070
2200	142·558	4700	304·555	7200	466·553	9700	628·550
2300	149·038	4800	311·035	7300	473·033	9800	635·030
2400	155·518	4900	317·515	7400	479·513	9900	641·510
2500	161·998	5000	323·995	7500	485·993	10000	647·990
2600	168·477	5100	330·475	7600	492·472	11000	712·789
2700	174·957	5200	336·955	7700	498·952	12000	777·588
2800	181·437	5300	343·435	7800	505·432	13000	842·387
2900	187·917	5400	349·915	7900	511·912	14000	907·186
3000	194·397	5500	356·395	8000	518·392	15000	971·985
3100	200·877	5600	362·874	8100	524·872	16000	1036·784
3200	207·357	5700	369·354	8200	531·352	17000	1101·583
3300	213·837	5800	375·834	8300	537·832	18000	1166·382
3400	220·317	5900	382·314	8400	544·312	19000	1231·181
3500	226·797	6000	388·794	8500	550·792	20000	1295·980

Grams
to grains

g	gr	g	gr	g	gr	g	gr
1	15·432	26	401·241	51	787·050	76	1172·859
2	30·865	27	416·674	52	802·483	77	1188·292
3	46·297	28	432·106	53	817·915	78	1203·724
4	61·729	29	447·538	54	833·347	79	1219·156
5	77·162	30	462·971	55	848·780	80	1234·589
6	92·594	31	478·403	56	864·212	81	1250·021
7	108·027	32	493·835	57	879·644	82	1265·454
8	123·459	33	509·268	58	895·077	83	1280·886
9	138·891	34	524·700	59	910·507	84	1296·318
10	154·324	35	540·133	60	925·942	85	1311·751
11	169·756	36	555·565	61	941·374	86	1327·183
12	185·188	37	570·997	62	956·806	87	1342·615
13	200·621	38	586·430	63	972·239	88	1358·048
14	216·053	39	601·862	64	987·671	89	1373·480
15	231·485	40	617·294	65	1003·103	90	1388·912
16	246·918	41	632·727	66	1018·536	91	1404·345
17	262·350	42	648·159	67	1033·968	92	1419·777
18	277·782	43	663·591	68	1049·400	93	1435·209
19	293·215	44	679·024	69	1064·833	94	1450·642
20	308·647	45	694·456	70	1080·265	95	1466·074
21	324·080	46	709·888	71	1095·698	96	1481·507
22	339·512	47	725·321	72	1111·130	97	1496·939
23	354·944	48	740·753	73	1126·562	98	1512·371
24	370·377	49	756·186	74	1141·995	99	1527·804
25	385·809	50	771·618	75	1157·427	100	1543·236

Grams
to grains

g	gr	g	gr	g	gr	g	gr
110	1697·564	360	5555·664	610	9413·764	860	13271·86
120	1851·888	370	5709·988	620	9568·088	870	13426·19
130	2006·212	380	5864·312	630	9722·412	880	13580·51
140	2160·536	390	6018·636	640	9876·736	890	13734·84
150	2314·860	400	6172·960	650	10031·060	900	13889·16
160	2469·184	410	6327·284	660	10185·384	910	14043·48
170	2623·508	420	6481·608	670	10339·708	920	14197·81
180	2777·832	430	6635·932	680	10494·032	930	14352·13
190	2932·156	440	6790·256	690	10648·356	940	14506·46
200	3086·480	450	6944·580	700	10802·680	950	14660·78
210	3240·804	460	7098·904	710	10957·004	960	14815·10
220	3395·128	470	7253·228	720	11111·328	970	14969·43
230	3549·452	480	7407·552	730	11265·652	980	15123·75
240	3703·776	490	7561·876	740	11419·976	990	15278·08
250	3858·100	500	7716·200	750	11574·300	1000	15432·40
260	4012·424	510	7870·524	760	11728·624	2000	30864·80
270	4166·748	520	8024·848	770	11882·948	3000	46297·20
280	4321·072	530	8179·172	780	12037·272	4000	61729·60
290	4475·396	540	8333·496	790	12191·596	5000	77162·00
300	4629·720	550	8487·820	800	12345·920	6000	92594·40
310	4784·044	560	8642·144	810	12500·244	7000	108026·80
320	4938·368	570	8796·468	820	12654·568	8000	123459·20
330	5092·692	580	8950·792	830	12808·892	9000	138891·60
340	5247·016	590	9105·116	840	12963·216	10000	154324·00
350	5401·340	600	9259·440	850	13117·540	25000	385810·00

Ounces (avoirdupois) to grams

oz	g	oz	g	oz	g	oz	g
1	28·35	26	737·09	51	1445·83	76	2154·56
2	56·70	27	765·44	52	1474·18	77	2182·91
3	85·05	28	793·79	53	1502·52	78	2211·26
4	113·40	29	822·14	54	1530·87	79	2239·61
5	141·75	30	850·49	55	1559·22	80	2267·96
6	170·10	31	878·84	56	1587·57	81	2296·31
7	198·45	32	907·19	57	1615·92	82	2324·66
8	226·80	33	935·53	58	1644·27	83	2353·01
9	255·15	34	963·88	59	1672·62	84	2381·36
10	283·50	35	992·23	60	1700·97	85	2409·71
11	311·86	36	1020·58	61	1729·32	86	2438·06
12	340·19	37	1048·93	62	1757·67	87	2466·41
13	368·54	38	1077·28	63	1786·02	88	2494·76
14	396·89	39	1105·63	64	1814·37	89	2523·11
15	425·24	40	1133·98	65	1842·72	90	2551·46
16	453·59	41	1162·33	66	1871·07	91	2579·81
17	481·94	42	1190·68	67	1899·42	92	2608·16
18	510·29	43	1219·03	68	1927·77	93	2636·51
19	538·64	44	1247·38	69	1956·12	94	2664·86
20	566·99	45	1275·73	70	1984·47	95	2693·20
21	595·34	46	1304·08	71	2012·82	96	2721·55
22	623·69	47	1322·43	72	2041·17	97	2749·90
23	652·04	48	1360·78	73	2069·52	98	2778·25
24	680·39	49	1389·13	74	2097·86	99	2806·60
25	708·74	50	1417·48	75	2126·21	100	2834·95

oz	g	oz	g	oz	g	oz	g
101	2863·30	126	3572·04	151	4280·77	176	4989·51
102	2891·65	127	3600·39	152	4309·12	177	5017·86
103	2920·00	128	3628·74	153	4337·47	178	5046·21
104	2948·35	129	3657·09	154	4365·82	179	5074·56
105	2976·70	130	3685·44	155	4394·17	180	5102·91
106	3005·05	131	3713·78	156	4422·52	181	5131·26
107	3033·40	132	3742·13	157	4450·87	182	5159·61
108	3061·75	133	3770·48	158	4479·22	183	5187·96
109	3090·10	134	3798·83	159	4507·57	184	5216·31
110	3118·45	135	3827·18	160	4535·92	185	5244·66
111	3146·79	136	3855·53	161	4564·27	186	5273·01
112	3175·14	137	3883·88	162	4592·62	187	5301·36
113	3203·49	138	3912·23	163	4620·97	188	5329·71
114	3231·84	139	3940·58	164	4649·32	189	5358·06
115	3260·19	140	3968·93	165	4677·67	190	5386·41
116	3288·54	141	3997·28	166	4706·02	191	5414·75
117	3316·89	142	4025·63	167	4734·37	192	5443·10
118	3345·24	143	4053·98	168	4762·72	193	5471·45
119	3373·59	144	4082·33	169	4791·07	194	5499·80
120	3401·94	145	4110·68	170	4819·42	195	5528·15
121	3430·29	146	4139·03	171	4847·76	196	5556·50
122	3458·64	147	4167·38	172	4876·11	197	5584·85
123	3486·99	148	4195·73	173	4904·46	198	5613·20
124	3515·34	149	4224·08	174	4932·81	199	5641·55
125	3543·69	150	4252·43	175	4961·16	200	5669·90

Grams
to ounces (avoirdupois)

g	oz	g	oz	g	oz	g	oz
1	0·0353	26	0·9171	51	1·7990	76	2·6808
2	0·0705	27	0·9524	52	1·8343	77	2·7161
3	0·1058	28	0·9877	53	1·8695	78	2·7514
4	0·1411	29	1·0229	54	1·9048	79	2·7866
5	0·1764	30	1·0582	55	1·9401	80	2·8219
6	0·2116	31	1·0935	56	1·9753	81	2·8572
7	0·2469	32	1·1288	57	2·0106	82	2·8925
8	0·2822	33	1·1640	58	2·0459	83	2·9277
9	0·3175	34	1·1993	59	2·0812	84	2·9630
10	0·3527	35	1·2346	60	2·1164	85	2·9983
11	0·3880	36	1·2699	61	2·1517	86	3·0336
12	0·4233	37	1·3051	62	2·1870	87	3·0688
13	0·4586	38	1·3404	63	2·2223	88	3·1041
14	0·4938	39	1·3757	64	2·2575	89	3·1394
15	0·5291	40	1·4110	65	2·2928	90	3·1747
16	0·5644	41	1·4462	66	2·3281	91	3·2099
17	0·5997	42	1·4815	67	2·3634	92	3·2452
18	0·6349	43	1·5168	68	2·3986	93	3·2805
19	0·6702	44	1·5521	69	2·4339	94	3·3158
20	0·7056	45	1·5873	70	2·4692	95	3·3510
21	0·7408	46	1·6226	71	2·5045	96	3·3863
22	0·7760	47	1·6579	72	2·5397	97	3·4216
23	0·8113	48	1·6932	73	2·5750	98	3·4569
24	0·8466	49	1·7284	74	2·6103	99	3·4921
25	0·8819	50	1·7637	75	2·6456	100	3·5274

Grams
to ounces (avoirdupois)

g	oz	g	oz	g	oz	g	oz
110	3·8801	360	12·6986	610	21·5171	860	30·3356
120	4·2329	370	13·0514	620	21·8699	870	30·6884
130	4·5856	380	13·4041	630	22·2226	880	31·0411
140	4·9384	390	13·7569	640	22·5754	890	31·3939
150	5·2911	400	14·1096	650	22·9281	900	31·7466
160	5·6438	410	14·4623	660	23·2808	910	32·0993
170	5·9966	420	14·8151	670	23·6336	920	32·4521
180	6·3493	430	15·1678	680	23·9863	930	32·8048
190	6·7021	440	15·5206	890	24·3391	940	33·1576
200	7·0548	450	15·8733	700	24·6918	950	33·5103
210	7·4075	460	16·2260	710	25·0445	960	33·8630
220	7·7603	470	16·5788	720	25·3973	970	34·2158
230	8·1130	480	16·9315	730	25·7500	980	34·5685
240	8·4658	490	17·2843	740	26·1028	990	34·9213
250	8·8185	500	17·6370	750	26·4555	1000	35·2740
260	9·1712	510	17·9897	760	26·8082	2000	70·5480
270	9·5240	520	18·3425	770	27·1610	3000	105·8220
280	9·8767	530	18·6952	780	27·5137	4000	141·0960
290	10·2295	540	19·0480	790	27·8665	5000	176·3700
300	10·5822	550	19·4007	800	28·2192	6000	211·6440
310	10·9349	560	19·7534	810	28·5719	7000	246·9180
320	11·2877	570	20·1062	820	28·9247	8000	282·1920
330	11·6404	580	20·4589	830	29·2774	9000	317·4660
340	11·9932	590	20·8117	840	29·6302	10000	352·7400
350	12·3459	600	21·1644	850	29·9829	25000	881·8500

Pounds
to kilograms

lb	kg	lb	kg	lb	kg	lb	kg
1	0·4536	26	11·7934	51	23·1332	76	34·4730
2	0·9072	27	12·2470	52	23·5868	77	34·9266
3	1·3608	28	12·7006	53	24·0404	78	35·3802
4	1·8144	29	13·1542	54	24·4940	79	35·8338
5	2·2680	30	13·6078	55	24·9476	80	36·2874
6	2·7216	31	14·0614	56	25·4012	81	36·7410
7	3·1752	32	14·5150	57	25·8548	82	37·1946
8	3·6287	33	14·9685	58	26·3084	83	37·6482
9	4·0823	34	15·4221	59	26·7619	84	38·1018
10	4·5359	35	15·8757	60	27·2155	85	38·5554
11	4·9895	36	16·3293	61	27·6691	86	39·0089
12	5·4431	37	16·7829	62	28·1227	87	39·4625
13	5·8967	38	17·2365	63	28·5763	88	39·9161
14	6·3503	39	17·6901	64	29·0299	89	40·3697
15	6·8039	40	18·1437	65	29·4835	90	40·8233
16	7·2575	41	18·5973	66	29·9371	91	41·2769
17	7·7111	42	19·0509	67	30·3907	92	41·7305
18	8·1647	43	19·5045	68	30·8443	93	42·1841
19	8·6183	44	19·9581	69	31·2979	94	42·6377
20	9·0719	45	20·4117	70	31·7515	95	43·0913
21	9·5254	46	20·8652	71	32·2051	96	43·5449
22	9·9790	47	21·3188	72	32·6587	97	43·9985
23	10·4326	48	21·7724	73	33·1122	98	44·4521
24	10·8862	49	22·2260	74	33·5658	99	44·9056
25	11·3398	50	22·6796	75	34·0194	100	45·3592

Pounds
to kilograms

lb	kg	lb	kg	lb	kg	lb	kg
110	49·895	360	163·293	610	276·691	860	390·089
120	54·431	370	167·829	620	281·227	870	394·625
130	58·967	380	172·365	630	285·763	880	399·161
140	63·503	390	176·901	640	290·299	890	403·697
150	68·039	400	181·437	650	294·835	900	408·233
160	72·575	410	185·973	660	299·371	910	412·769
170	77·111	420	190·509	670	303·907	920	417·305
180	81·647	430	195·045	680	308·443	930	421·841
190	86·182	440	199·580	690	312·978	940	426·376
200	90·718	450	204·116	700	317·514	950	430·912
210	95·254	460	208·652	710	322·050	960	435·448
220	99·790	470	213·188	720	326·586	970	439·984
230	104·326	480	217·724	730	331·122	980	444·520
240	108·862	490	222·260	740	335·658	990	449·056
250	113·398	500	226·796	750	340·194	1000	453·592
260	117·934	510	231·332	760	344·730	2000	907·184
270	122·470	520	235·868	770	349·266	3000	1360·776
280	127·006	530	240·404	780	353·802	4000	1814·368
290	131·542	540	244·940	790	358·338	5000	2267·960
300	136·078	550	249·476	800	362·874	6000	2721·552
310	140·614	560	254·012	810	367·410	7000	3175·144
320	145·149	570	258·547	820	371·945	8000	3628·736
330	149·685	580	263·083	830	376·481	9000	4082·328
340	154·221	590	267·619	840	381·017	10000	4535·920
350	158·757	600	272·155	850	385·553	25000	11339·800

Kilograms
to pounds

kg	lb	kg	lb	kg	lb	kg	lb
1	2·205	26	57·320	51	112·436	76	167·551
2	4·409	27	59·525	52	114·640	77	169·756
3	6·614	28	61·729	53	116·845	78	171·961
4	8·819	29	63·934	54	119·050	79	174·165
5	11·023	30	66·139	55	121·254	80	176·370
6	13·228	31	68·343	56	123·459	81	178·574
7	15·432	32	70·548	57	125·663	82	180·779
8	17·637	33	72·753	58	127·868	83	182·984
9	19·842	34	74·957	59	130·073	84	185·188
10	22·046	35	77·162	60	132·277	85	187·393
11	24·251	36	79·366	61	134·482	86	189·598
12	26·456	37	81·571	62	136·687	87	191·802
13	28·660	38	83·776	63	138·891	88	194·007
14	30·865	39	85·980	64	141·096	89	196·211
15	33·069	40	88·185	65	143·300	90	198·416
16	35·274	41	90·390	66	145·505	91	200·621
17	37·479	42	92·594	67	147·710	92	202·825
18	39·683	43	94·799	68	149·914	93	205·030
19	41·888	44	97·003	69	152·119	94	207·235
20	44·093	45	99·208	70	154·324	95	209·439
21	46·297	46	101·413	71	156·528	96	211·644
22	48·502	47	103·617	72	158·733	97	213·848
23	50·706	48	105·822	73	160·937	98	216·053
24	52·911	49	108·027	74	163·142	99	218·258
25	55·116	50	110·231	75	165·347	100	220·462

Kilograms to pounds

kg	lb	kg	lb	kg	lb	kg	lb
110	242·508	360	793·663	610	1344·818	860	1895·973
120	264·554	370	815·709	620	1366·864	870	1918·019
130	286·601	380	837·756	630	1388·911	880	1940·066
140	308·647	390	859·802	640	1410·957	890	1962·112
150	330·693	400	881·848	650	1433·003	900	1984·158
160	352·739	410	903·894	660	1455·049	910	2006·204
170	374·785	420	925·940	670	1477·095	920	2028·250
180	396·832	430	947·987	680	1499·142	930	2050·297
190	418·878	440	970·033	690	1521·188	940	2072·343
200	440·924	450	992·079	700	1543·234	950	2094·389
210	462·970	460	1014·125	710	1565·280	960	2116·435
220	485·016	470	1036·171	720	1587·326	970	2138·481
230	507·063	480	1058·218	730	1609·373	980	2160·528
240	529·109	490	1080·264	740	1631·419	990	2182·574
250	551·155	500	1102·310	750	1653·465	1000	2204·620
260	573·201	510	1124·356	760	1675·511	2000	4409·240
270	595·247	520	1146·402	770	1697·557	3000	6613·860
280	617·294	530	1168·449	780	1719·604	4000	8818·480
290	639·340	540	1190·495	790	1741·650	5000	11023·100
300	661·386	550	1212·541	800	1763·696	6000	13227·720
310	683·432	560	1234·587	810	1785·742	7000	15432·340
320	705·478	570	1256·633	820	1807·788	8000	17636·960
330	727·525	580	1278·680	830	1829·835	9000	19841·580
340	749·571	590	1300·726	840	1851·881	10000	22046·200
350	771·617	600	1322·772	850	1873·927	25000	55115·500

UK hundredweights to kilograms

UK cwt	kg	UK cwt	kg	UK cwt	kg	UK cwt	kg
1	50·802	26	1320·860	51	2590·917	76	3860·975
2	101·605	27	1371·662	52	2641·720	77	3911·777
3	152·407	28	1422·464	53	2692·522	78	1962·579
4	203·209	29	1473·267	54	2743·324	79	4013·382
5	254·012	30	1524·069	55	2794·127	80	4064·184
6	304·814	31	1574·871	56	2844·929	81	4114·986
7	355·616	32	1625·674	57	2895·731	82	4165·789
8	406·418	33	1676·476	58	2946·533	83	4216·591
9	457·221	34	1727·278	59	2997·336	84	4267·393
10	508·023	35	1778·081	60	3048·138	85	4318·196
11	558·825	36	1828·883	61	3098·940	86	4368·998
12	609·628	37	1879·685	62	3149·743	87	4419·800
13	660·430	38	1930·487	63	3200·545	88	4470·602
14	711·232	39	1981·290	64	3251·347	89	4521·405
15	762·035	40	2032·092	65	3302·150	90	4572·207
16	812·837	41	2082·894	66	3352·952	91	4623·009
17	863·639	42	2133·697	67	3403·754	92	4673·812
18	914·441	43	2184·499	68	3454·556	93	4724·614
19	965·244	44	2235·301	69	3505·359	94	4775·416
20	1016·046	45	2286·104	70	3556·161	95	4826·219
21	1066·848	46	2336·906	71	3606·963	96	4877·021
22	1117·651	47	2387·708	72	3657·766	97	4927·823
23	1168·453	48	2438·510	73	3708·568	98	4978·625
24	1219·255	49	2489·313	74	3759·370	99	5029·428
25	1270·058	50	2540·115	75	3810·173	100	5080·230

Kilograms
to UK hundredweights

kg	UK cwt	kg	UK cwt	kg	UK cwt	kg	UK cwt
100	1·9684	350	6·8894	600	11·8104	850	16·7314
110	2·1652	360	7·0862	610	12·0072	860	16·9282
120	2·3621	370	7·2831	620	12·2041	870	17·1251
130	2·5589	380	7·4799	630	12·4009	880	17·3219
140	2·7558	390	7·6768	640	12·5978	890	17·5188
150	2·9526	400	7·8736	650	12·7946	900	17·7156
160	3·1494	410	8·0704	660	12·9914	910	17·9124
170	3·3463	420	8·2673	670	13·1883	920	18·1093
180	3·5431	430	8·4641	680	13·3851	930	18·3061
190	3·7400	440	8·6610	690	13·5820	940	18·5030
200	3·9368	450	8·8578	700	13·7788	950	18·6998
210	4·1336	460	9·0546	710	13·9756	960	18·8966
220	4·3305	470	9·2515	720	14·1725	970	19·0935
230	4·5273	480	9·4483	730	14·3693	980	19·2903
240	4·7242	490	9·6452	740	14·5662	990	19·4872
250	4·9210	500	9·8420	750	14·7630	1000	19·6840
260	5·1178	510	10·0388	760	14·9598	2000	39·3680
270	5·3147	520	10·2357	770	15·1567	3000	59·0520
280	5·5115	530	10·4325	780	15·3535	4000	78·7360
290	5·7084	540	10·6294	790	15·5504	5000	98·4200
300	5·9052	550	10·8262	800	15·7472	6000	118·1040
310	6·1020	560	11·0230	810	15·9440	7000	137·7880
320	6·2989	570	11·2199	820	16·1409	8000	157·4720
330	6·4957	580	11·4167	830	16·3377	9000	177·1560
340	6·6926	590	11·6136	840	16·5346	10000	196·8400

Short (US) hundredweights to kilograms

sh cwt	kg	sh cwt	kg	sh cwt	kg	sh cwt	kg
1	45·359	26	1179·339	51	2313·319	76	3447·299
2	90·718	27	1224·698	52	2358·678	77	3492·658
3	136·078	28	1270·058	53	2404·038	78	3538·018
4	181·437	29	1315·417	54	2449·397	79	3583·377
5	226·796	30	1360·776	55	2494·756	80	3628·736
6	272·155	31	1406·135	56	2540·115	81	3674·095
7	317·514	32	1451·494	57	2585·474	82	3719·454
8	362·874	33	1496·854	58	2630·834	83	3764·814
9	408·233	34	1542·213	59	2676·193	84	3810·173
10	453·592	35	1587·572	60	2721·552	85	3855·532
11	498·951	36	1632·931	61	2766·911	86	3900·891
12	544·310	37	1678·290	62	2812·270	87	3946·250
13	589·670	38	1723·650	63	2857·630	88	3991·610
14	635·029	39	1769·009	64	2902·989	89	4036·969
15	680·388	40	1814·368	65	2948·348	90	4082·328
16	725·747	41	1859·727	66	2993·707	91	4127·687
17	771·106	42	1905·086	67	3039·066	92	4173·046
18	816·466	43	1950·446	68	3084·426	93	4218·406
19	861·825	44	1995·805	69	3129·785	94	4263·765
20	907·184	45	2041·164	70	3175·144	95	4309·124
21	952·543	46	2086·523	71	3220·503	96	4354·483
22	997·902	47	2131·882	72	3265·862	97	4399·842
23	1043·262	48	2177·242	73	3311·222	98	4445·202
24	1088·621	49	2222·601	74	3356·581	99	4490·561
25	1133·980	50	2267·960	75	3401·940	100	4535·920

Kilograms
to short (US) hundredweights

kg	sh cwt	kg	sh cwt	kg	sh cwt	kg	sh cwt
100	2·2046	350	7·7161	600	13·2276	850	18·7391
110	2·4251	360	7·9366	610	13·4481	860	18·9596
120	2·6455	370	8·1570	620	13·6685	870	19·1800
130	2·8660	380	8·3775	630	13·8890	880	19·4005
140	3·0864	390	8·5979	640	14·1094	890	19·6209
150	3·3069	400	8·8184	650	14·3299	900	19·8414
160	3·5274	410	9·0389	660	14·5504	910	20·0619
170	3·7478	420	9·2593	670	14·7708	920	20·2823
180	3·9683	430	9·4798	680	14·9913	930	20·5028
190	4·1887	440	9·7002	690	15·2117	940	20·7232
200	4·4092	450	9·9207	700	15·4322	950	20·9437
210	4·6297	460	10·1412	710	15·6527	960	21·1642
220	4·8501	470	10·3616	720	15·8731	970	21·3846
230	5·0706	480	10·5821	730	16·0936	980	21·6051
240	5·2910	490	10·8025	740	16·3140	990	21·8255
250	5·5115	500	11·0230	750	16·5345	1000	22·0460
260	5·7320	510	11·2435	760	16·7550	2000	44·0920
270	5·9524	520	11·4639	770	16·9754	3000	66·1380
280	6·1729	530	11·6844	780	17·1959	4000	88·1840
290	6·3933	540	11·9048	790	17·4163	5000	110·2300
300	6·6138	550	12·1253	800	17·6368	6000	132·2760
310	6·8343	560	12·3458	810	17·8573	7000	154·3220
320	7·0547	570	12·5662	820	18·0777	8000	176·3680
330	7·2752	580	12·7867	830	18·2982	9000	198·4140
340	7·4956	590	13·0071	840	18·5186	10000	220·4600

UK tons
to tonnes

UK tons	tonnes	UK tons	tonnes	UK tons	tonnes	UK tons	tonnes
1	1·0160	26	26·4172	51	51·8184	76	77·2196
2	2·0321	27	27·4333	52	52·8344	77	78·2356
3	3·0481	28	28·4493	53	53·8505	78	79·2517
4	4·0642	29	29·4654	54	54·8665	79	80·2677
5	5·0802	30	30·4814	55	55·8826	80	81·2838
6	6·0963	31	31·4975	56	56·8986	81	82·2998
7	7·1123	32	32·5135	57	57·9147	82	83·3158
8	8·1284	33	33·5295	58	58·9307	83	84·3319
9	9·1444	34	34·5456	59	59·9468	84	85·3479
10	10·1605	35	35·5616	60	60·9628	85	86·3640
11	11·1765	36	36·5777	61	61·9789	86	87·3800
12	12·1926	37	37·5937	62	62·9949	87	88·3961
13	13·2086	38	38·6098	63	64·0110	88	89·4121
14	14·2247	39	39·6258	64	65·0270	89	90·4282
15	15·2407	40	40·6419	65	66·0430	90	91·4442
16	16·2568	41	41·6579	66	67·0591	91	92·4603
17	17·2728	42	42·6740	67	68·0751	92	93·4763
18	18·2888	43	43·6900	68	69·0912	93	94·4924
19	19·3049	44	44·7061	69	70·1072	94	95·5084
20	20·3209	45	45·7221	70	71·1233	95	96·5245
21	21·3370	46	46·7382	71	72·1393	96	97·5405
22	22·3530	47	47·7542	72	73·1554	97	98·5566
23	23·3691	48	48·7703	73	74·1714	98	99·5726
24	24·3851	49	49·7863	74	75·1875	99	100·5886
25	25·4012	50	50·8023	75	76·2035	100	101·6047

UK tons	tonnes	UK tons	tonnes	UK tons	tonnes	UK tons	tonnes
110	111·766	360	365·778	610	619·791	860	873·803
120	121·926	370	375·939	620	629·951	870	883·964
130	132·087	380	386·099	630	640·112	880	894·124
140	142·247	390	396·260	640	650·272	890	904·285
150	152·408	400	406·420	650	660·433	900	914·445
160	162·568	410	416·581	660	670·593	910	924·606
170	172·729	420	426·741	670	680·754	920	934·766
180	182·889	430	436·902	680	690·914	930	944·927
190	193·050	440	447·062	690	701·075	940	955·087
200	203·210	450	457·223	700	711·235	950	965·248
210	213·371	460	467·383	710	721·396	960	975·408
220	223·531	470	477·544	720	731·556	970	985·569
230	233·692	480	487·704	730	741·717	980	995·729
240	243·852	490	497·865	740	751·877	990	1005·890
250	254·013	500	508·025	750	762·038	1000	1016·050
260	264·173	510	518·186	760	772·198	2000	2032·100
270	274·334	520	528·346	770	782·359	3000	3048·150
280	284·494	530	538·507	780	792·519	4000	4064·200
290	294·655	540	548·667	790	802·680	5000	5080·250
300	304·815	550	558·828	800	812·840	6000	6096·300
310	314·976	560	568·988	810	823·001	7000	7112·350
320	325·136	570	579·149	820	833·161	8000	8128·400
330	335·297	580	589·309	830	843·322	9000	9144·450
340	345·457	590	599·470	840	853·482	10000	10160·500
350	355·618	600	609·630	850	863·643	25000	25401·250

Tonnes
to UK tons

tonnes	UK tons	tonnes	UK tons	tonnes	UK tons	tonnes	UK tons
1	0·9842	26	25·5894	51	50·1945	76	74·7997
2	1·9684	27	26·5736	52	51·1787	77	75·7839
3	2·9526	28	27·5578	53	52·1629	78	76·7681
4	3·9368	29	28·5420	54	53·1472	79	77·7523
5	4·9210	30	29·5262	55	54·1314	80	78·7365
6	5·9052	31	30·5104	56	55·1156	81	79·7207
7	6·8894	32	31·4946	57	56·0998	82	80·7049
8	7·8737	33	32·4788	58	57·0840	83	81·6891
9	8·8579	34	33·4630	59	58·0682	84	82·6733
10	9·8421	35	34·4472	60	59·0524	85	83·6576
11	10·8263	36	35·4314	61	60·0366	86	84·6418
12	11·8105	37	36·4156	62	61·0208	87	85·6260
13	12·7947	38	37·3998	63	62·0050	88	86·6102
14	13·7789	39	38·3841	64	62·9892	89	87·5944
15	14·7631	40	39·3683	65	63·9734	90	88·5786
16	15·7473	41	40·3525	66	64·9576	91	89·5628
17	16·7315	42	41·3367	67	65·9418	92	90·5470
18	17·7157	43	42·3209	68	66·9260	93	91·5312
19	18·6999	44	43·3051	69	67·9103	94	92·5154
20	19·6841	45	44·2893	70	68·8945	95	93·4996
21	20·6683	46	45·2735	71	69·8787	96	94·4838
22	21·6525	47	46·2577	72	70·8629	97	95·4680
23	22·6368	48	47·2419	73	71·8471	98	96·4522
24	23·6210	49	48·2261	74	72·8313	99	97·4364
25	24·6052	50	49·2103	75	73·8155	100	98·4207

Tonnes
to UK tons

tonnes	UK tons	tonnes	UK tons	tonnes	UK tons	tonnes	UK tons
110	108·263	360	354·315	610	600·366	860	846·418
120	118·105	370	364·157	620	610·208	870	856·260
130	127·947	380	373·999	630	620·050	880	866·102
140	137·789	390	383·841	640	629·892	890	875·944
150	147·631	400	393·683	650	639·735	900	885·786
160	157·473	410	403·525	660	649·577	910	895·628
170	167·315	420	413·367	670	659·419	920	905·470
180	177·157	430	423·209	680	669·261	930	915·313
190	186·999	440	433·051	690	679·103	940	925·155
200	196·841	450	442·893	700	688·945	950	934·997
210	206·683	460	452·735	710	698·787	960	944·839
220	216·526	470	462·577	720	708·629	970	954·681
230	226·368	480	472·419	730	718·471	980	964·523
240	236·210	490	482·261	740	728·313	990	974·365
250	246·052	500	492·104	750	738·155	1000	984·207
260	255·894	510	501·946	760	747·997	2000	1968·414
270	265·736	520	511·788	770	757·839	3000	2952·621
280	275·578	530	521·630	780	767·681	4000	3936·828
290	285·420	540	531·472	790	777·524	5000	4921·035
300	295·262	550	541·314	800	787·366	6000	5905·242
310	305·104	560	551·156	810	797·208	7000	6889·449
320	314·946	570	560·998	820	807·050	8000	7873·656
330	324·788	580	570·840	830	816·892	9000	8857·863
340	334·630	590	580·682	840	826·734	10000	9842·070
350	344·472	600	590·524	850	836·576	25000	24605·175

Short (US) tons to tonnes

sh tons	tonnes	sh tons	tonnes	sh tons	tonnes	sh tons	tonnes
1	0·9072	26	23·5868	51	46·2664	76	68·9461
2	1·8144	27	24·4940	52	47·1736	77	69·8532
3	2·7216	28	25·4012	53	48·0808	78	70·7604
4	3·6287	29	26·3084	54	48·9880	79	71·6676
5	4·5359	30	27·2156	55	49·8952	80	72·5748
6	5·4431	31	28·1227	56	50·8024	81	73·4820
7	6·3503	32	29·0299	57	51·7095	82	74·3892
8	7·2575	33	29·9371	58	52·6167	83	75·2964
9	8·1647	34	30·8443	59	53·5239	84	76·2035
10	9·0719	35	31·7515	60	54·4311	85	77·1107
11	9·9790	36	32·6587	61	55·3383	86	78·0179
12	10.8862	37	33·5658	62	56·2455	87	78·9251
13	11·7934	38	34·4730	63	57·1527	88	79·8323
14	12·7006	39	35·3802	64	58·0598	89	80·7395
15	13·6078	40	36·2874	65	58·9670	90	81·6467
16	14·5150	41	37·1946	66	59·8742	91	82·5538
17	15·4221	42	38·1018	67	60·7814	92	83·4610
18	16·3293	43	39·0090	68	61·6886	93	84·3682
19	17·2365	44	39·9161	69	62·5958	94	85·2754
20	18·1437	45	40·8233	70	63·5030	95	86·1826
21	19·0509	46	41·7305	71	64·4101	96	87·0898
22	19·9581	47	42·6377	72	65·3173	97	87·9969
23	20·8653	48	43·5449	73	66·2245	98	88·9041
24	21·7724	49	44·4521	74	67·1317	99	89·8113
25	22·6796	50	45·3593	75	68·0389	100	90·7185

Short (US) tons to tonnes

sh tons	tonnes	sh tons	tonnes	sh tons	tonnes	sh tons	tonnes
110	99·790	360	326·587	610	553·383	860	780·179
120	108·862	370	335·658	620	562·455	870	789·251
130	117·934	380	344·730	630	571·527	880	798·323
140	127·006	390	353·802	640	580·598	890	807·395
150	136·078	400	362·874	650	589·670	900	816·467
160	145·150	410	371·946	660	598·742	910	825·538
170	154·221	420	381·018	670	607·814	920	834·610
180	163·293	430	390·090	680	616·886	930	843·682
190	172·365	440	399·161	690	625·958	940	852·754
200	181·437	450	408·233	700	635·030	950	861·826
210	190·509	460	417·305	710	644·101	960	870·898
220	199·581	470	426·377	720	653·173	970	879·969
230	208·653	480	435·449	730	662·245	980	889·041
240	217·724	490	444·521	740	671·317	990	898·113
250	226·796	500	453·593	750	680·389	1000	907·185
260	235·868	510	462·664	760	689·461	2000	1814·370
270	244·940	520	471·736	770	698·532	3000	2721·555
280	254·012	530	480·808	780	707·604	4000	3628·740
290	263·084	540	489·880	790	716·676	5000	4535·925
300	272·156	550	498·952	800	725·748	6000	5443·110
310	281·227	560	508·024	810	734·820	7000	6350·295
320	290·299	570	517·095	820	743·892	8000	7257·480
330	299·371	580	526·167	830	752·964	9000	8164·665
340	308·443	590	535·239	840	762·035	10000	9071·850
350	317·515	600	544·311	850	771·107	25000	22679·625

Tonnes
to short (US) tons

tonnes	sh tons	tonnes	sh tons	tonnes	sh tons	tonnes	sh tons
1	1·1023	26	28·6601	51	56·2178	76	83·7756
2	2·2046	27	29·7624	52	57·3201	77	84·8779
3	3·3069	28	30·8647	53	58·4224	78	85·9802
4	4·4092	29	31·9670	54	59·5247	79	87·0825
5	5·5116	30	33·0693	55	60·6271	80	88·1848
6	6·6139	31	34·1716	56	61·7294	81	89·2871
7	7·7162	32	35·2739	57	62·8317	82	90·3894
8	8·8185	33	36·3762	58	63·9340	83	91·4917
9	9·9208	34	37·4785	59	65·0363	84	92·5940
10	11·0231	35	38·5809	60	66·1386	85	93·6964
11	12·1254	36	39·6832	61	67·2409	86	94·7987
12	13·2277	37	40·7855	62	68·3432	87	95·9010
13	14·3300	38	41·8878	63	69·4455	88	97·0033
14	15·4323	39	42·9901	64	70·5478	89	98·1056
15	16·5347	40	44·0924	65	71·6502	90	99·2079
16	17·6370	41	45·1947	66	72·7525	91	100·3102
17	18·7393	42	46·2970	67	73·8548	92	101·4125
18	19·8416	43	47·3993	68	74·9571	93	102·5148
19	20·9439	44	48·5016	69	76·0594	94	103·6171
20	22·0462	45	49·6040	70	77·1617	95	104·7195
21	23·1485	46	50·7063	71	78·2640	96	105·8218
22	24·2508	47	51·8086	72	79·3663	97	106·9241
23	25·3531	48	52·9109	73	80·4686	98	108·0264
24	26·4554	49	54·0132	74	81·5709	99	109·1287
25	27·5578	50	55·1155	75	82·6733	100	110·2310

Tonnes
to short (US) tons

tonnes	sh tons	tonnes	sh tons	tonnes	sh tons	tonnes	sh tons
110	121·254	360	396·832	610	672·409	860	947·987
120	132·277	370	407·855	620	683·432	870	959·010
130	143·300	380	418·878	630	694·455	880	970·033
140	154·323	390	429·901	640	705·478	890	981·056
150	165·347	400	440·924	650	716·502	900	992·079
160	176·370	410	451·947	660	727·525	910	1003·102
170	187·393	420	462·970	670	738·548	920	1014·125
180	198·416	430	473·993	680	749·571	930	1025·148
190	209·439	440	485·016	690	760·594	940	1036·171
200	220·462	450	496·040	700	771·617	950	1047·195
210	231·485	460	507·063	710	782·640	960	1058·218
220	242·508	470	518·086	720	793·663	970	1069·241
230	253·531	480	529·109	730	804·686	980	1080·264
240	264·554	490	540·132	740	815·709	990	1091·287
250	275·578	500	551·155	750	826·733	1000	1102·310
260	286·601	510	562·178	760	837·756	2000	2204·620
270	297·624	520	573·201	770	848·779	3000	3306·930
280	308·647	530	584·224	780	859·802	4000	4409·240
290	319·670	540	595·247	790	870·825	5000	5511·550
300	330·693	550	606·271	800	881·848	6000	6613·860
310	341·716	560	617·294	810	892·871	7000	7716·170
320	352·739	570	628·317	820	903·894	8000	8818·480
330	363·762	580	639·340	830	914·917	9000	9920·790
340	374·785	590	650·363	840	925·940	10000	11023·100
350	385·809	600	661·386	850	936·964	25000	27557·750

Pounds per square inch to kilograms per square centimetre

lb/in²	kg/cm²	lb/in²	kg/cm²	lb/in²	kg/cm²	lb/in²	kg/cm²
1	0·0703	26	1·8280	51	3·5857	76	5·3433
2	0·1406	27	1·8983	52	3·6560	77	5·4136
3	0·2109	28	1·9686	53	3·7263	78	5·4839
4	0·2812	29	2·0389	54	3·7966	79	5·5543
5	0·3515	30	2·1092	55	3·8669	80	5·6246
6	0·4218	31	2·1795	56	3·9372	81	5·6949
7	0·4921	32	2·2498	57	4·0075	82	5·7652
8	0·5625	33	2·3201	58	4·0778	83	5·8355
9	0·6328	34	2·3904	59	4·1481	84	5·9058
10	0·7031	35	2·4607	60	4·2184	85	5·9761
11	0·7734	36	2·5311	61	4·2887	86	6·0464
12	0·8437	37	2·6014	62	4·3590	87	6·1167
13	0·9140	38	2·6717	63	4·4293	88	6·1870
14	0·9843	39	2·7420	64	4·4996	89	6·2573
15	1·0546	40	2·8123	65	4·5700	90	6·3276
16	1·1249	41	2·8826	66	4·6403	91	6·3979
17	1·1952	42	2·9529	67	4·7106	92	6·4682
18	1·2655	43	3·0232	68	4·7809	93	6·5386
19	1·3358	44	3·0935	69	4·8512	94	6·6089
20	1·4061	45	3·1638	70	4·9215	95	6·6792
21	1·4764	46	3·2341	71	4·9918	96	6·7495
22	1·5468	47	3·3044	72	5·0621	97	6·8198
23	1·6171	48	3·3747	73	5·1324	98	6·8901
24	1·6874	49	3·4450	74	5·2027	99	6·9604
25	1·7577	50	3·5154	75	5·2730	100	7·0307

Pounds per square inch
to kilograms per square centimetre

lb/in²		lb/in²	kg/cm²	lb/in²	kg/cm²	lb/in²	kg/cm²
110	7·7338	360	25·3105	610	42·8873	860	60·4640
120	8·4368	370	26·0136	620	43·5903	870	61·1671
130	9·1399	380	26·7167	630	44·2934	880	61·8702
140	9·8430	390	27·4197	640	44·9965	890	62·5732
150	10·5461	400	28·1228	650	45·6996	900	63·2763
160	11·2491	410	28·8259	660	46·4026	910	63·9794
170	11·9522	420	29·5289	670	47·1057	920	64·6824
180	12·6553	430	30·2320	680	47·8088	930	65·3855
190	13·3583	440	30·9351	690	48·5118	940	66·0886
200	14·0614	450	31·6382	700	49·2149	950	66·7917
210	14·7645	460	32·3412	710	49·9180	960	67·4947
220	15·4675	470	33·0443	720	50·6210	970	68·1978
230	16·1706	480	33·7474	730	51·3241	980	68·9009
240	16·8737	490	34·4504	740	52·0272	990	69·6039
250	17·5768	500	35·1535	750	52·7303	1000	70·3070
260	18·2798	510	35·8566	760	53·4333	2000	140·6140
270	18·9829	520	36·5596	770	54·1364	3000	210·9210
280	19·6860	430	37·2627	780	54·8395	4000	281·2280
290	20·3890	540	37·9658	790	55·5425	5000	351·5350
300	21·0921	550	38·6689	800	56·2456	6000	421·8420
310	21·7952	560	39·3719	810	56·9487	7000	492·1490
320	22·4982	570	40·0750	820	57·6517	8000	562·4560
330	23·2013	580	40·7781	830	58·3548	9000	632·7630
340	23·9044	590	41·4811	840	59·0579	10000	703·0700
350	24·6075	600	42·1842	850	59·7610	15000	1054·6050

Kilograms per square centimetre to pounds per square inch

kg/cm²	lb/in²	kg/cm²	lb/in²	kg/cm²	lb/in²	kg/cm²	lb/in²
1	14·2233	26	369·8058	51	725·3883	76	1080·9708
2	28·4466	27	384·0291	52	739·6116	77	1095·1941
3	42·6699	28	398·2524	53	753·8349	78	1109·4174
4	56·8932	29	412·4757	54	768·0582	79	1123·6407
5	71·1165	30	426·6990	55	782·2815	80	1137·8640
6	85·3398	31	440·9223	56	796·5048	81	1152·0873
7	99·5631	32	455·1456	57	810·7281	82	1166·3106
8	113·7864	33	469·3689	58	824·9514	83	1180·5339
9	128·0097	34	483·5922	59	839·1747	84	1194·7572
10	142·2330	35	497·8155	60	853·3980	85	1208·9805
11	156·4563	36	512·0388	61	867·6213	86	1223·2038
12	170·6796	37	526·2621	62	881·8446	87	1237·4271
13	184·9029	38	540·4854	63	896·0679	88	1251·6504
14	199·1262	39	554·7087	64	910·2912	89	1265·8737
15	213·3495	40	568·9320	65	924·5145	90	1280·0970
16	227·5728	41	583·1553	66	938·7378	91	1294·3203
17	241·7961	42	597·3786	67	952·9611	92	1308·5436
18	256·0194	43	611·6019	68	967·1844	93	1322·7669
19	270·2427	44	625·8252	69	981·4077	94	1336·9902
20	284·4660	45	640·0485	70	995·6310	95	1351·2135
21	298·6893	46	654·2718	71	1009·8543	96	1365·4368
22	312·9126	47	668·4951	72	1024·0776	97	1379·6601
23	327·1359	48	682·7184	73	1038·3009	98	1393·8834
24	341·3592	49	696·9417	74	1052·5242	99	1408·1067
25	355·5825	50	711·1650	75	1066·7475	100	1422·3300

Kilograms per square centimetre to pounds per square inch

kg/cm^2	lb/in^2	kg/cm^2	lb/in^2	kg/cm^2	lb/in^2	kg/cm^2	lb/in^2
110	1564·563	360	5120·388	610	8676·213	860	12232·04
120	1706·796	370	5262·621	620	8818·446	870	12374·27
130	1849·029	380	5404·854	630	8960·679	880	12516·50
140	1991·262	390	5547·087	640	9102·912	890	12658·74
150	2133·495	400	5689·320	650	9245·145	900	12800·97
160	2275·728	410	5831·553	660	9387·378	910	12943·20
170	2417·961	420	5973·786	670	9529·611	920	13085·44
180	2560·194	430	6116·019	680	9671·844	930	13227·67
190	2702·427	440	6258·252	690	9814·077	940	13369·90
200	2844·660	450	6400·485	700	9956·310	950	13512·14
210	2986·893	460	6542·718	710	10098·543	960	13654·37
220	3129·126	470	6684·951	720	10240·776	970	13796·60
230	3271·359	480	6827·184	730	10383·009	980	13938·83
240	3413·592	490	6969·417	740	10525·242	990	14081·07
250	3555·825	500	7111·650	750	10667·475	1000	14223·30
260	3698·058	510	7253·883	760	10809·708	2000	28446·60
270	3840·291	520	7396·116	770	10951·941	3000	42669·90
280	3982·524	530	7538·349	780	11094·174	4000	56893·20
290	4124·757	540	7680·582	790	11236·407	5000	71116·50
300	4266·990	550	7822·815	800	11378·640	6000	85339·80
310	4409·223	560	7965·048	810	11520·873	7000	99563·10
320	4551·456	570	8107·281	820	11663·106	8000	113786·40
330	4693·689	580	8249·514	830	11805·339	9000	128009·70
340	4835·922	590	8391·747	840	11947·572	10000	142233·00
350	4978·155	600	8533·980	850	12089·805	15000	213349·50

Miles per UK gallon
to kilometres per litre

miles/ UK gal	km/l	miles/ UK gal	km/l	miles/ UK gal	km/l	miles/ UK gal	km/l
1	0·3540	26	9·2042	51	18·0543	76	26·9045
2	0·7080	27	9·5582	52	18·4083	77	27·2585
3	1·0620	28	9·9122	53	18·7623	78	27·6125
4	1·4160	29	10·2662	54	19·1163	79	27·9665
5	1·7700	30	10·6202	55	19·4703	80	28·3205
6	2·1240	31	10·9742	56	19·8243	81	28·6745
7	2·4780	32	11·3282	57	20·1783	82	29·0285
8	2·8320	33	11·6822	58	20·5323	83	29·3825
9	3·1861	34	12·0362	59	20·8864	84	29·7365
10	3·5401	35	12·3902	60	21·2404	85	30·0905
11	3·8941	36	12·7442	61	21·5944	86	30·4445
12	4·2481	37	13·0982	62	21·9484	87	30·7985
13	4·6021	38	13·4522	63	22·3024	88	31·1525
14	4·9561	39	13·8062	64	22·6564	89	31·5065
15	5·3101	40	14·1602	65	23·0104	90	31·8605
16	5·6641	41	14·5142	66	23·3644	91	32·2145
17	6·0181	42	14·8683	67	23·7184	92	32·5686
18	6·3721	43	15·2223	68	24·0724	93	32·9226
19	6·7261	44	15·5763	69	24·4264	94	33·2766
20	7·0801	45	15·9303	70	24·7804	95	33·6306
21	7·4341	46	16·2843	71	25·1344	96	33·9846
22	7·7881	47	16·6383	72	25·4884	97	34·3386
23	8·1421	48	16·9923	73	25·8424	98	34·6926
24	8·4961	49	17·3463	74	26·1964	99	35·0466
25	8·8502	50	17·7003	75	26·5505	100	35·4006

Kilometres per litre to miles per UK gallon

km/l	miles/ UK gal	km/l	miles/ UK gal	km/l	miles/ UK gal	km/l	miles/ UK gal
1	2·8248	26	73·4451	51	144·0653	76	214·6856
2	5·6496	27	76·2699	52	146·8901	77	217·5104
3	8·4744	28	79·0947	53	149·7149	78	220·3352
4	11·2992	29	81·9195	54	152·5397	79	223·1600
5	14·1241	30	84·7443	55	155·3646	80	225·9848
6	16·9489	31	87·5691	56	158·1894	81	228·8096
7	19·7737	32	90·3939	57	161·0142	82	231·6344
8	22·5985	33	93·2187	58	163·8390	83	234·4592
9	25·4233	34	96·0435	59	166·6638	84	237·2840
10	28·2481	35	98·8684	60	169·4886	85	240·1089
11	31·0729	36	101·6932	61	172·3134	86	242·9337
12	33·8977	37	104·5180	62	175·1382	87	245·7585
13	36·7225	38	107·3428	63	177·9630	88	248·5833
14	39·5473	39	110·1676	64	180·7878	89	251·4081
15	42·3722	40	112·9924	65	183·6127	90	254·2329
16	45·1970	41	115·8172	66	186·4375	91	257·0577
17	48·0218	42	118·6420	67	189·2623	92	259·8825
18	50·8466	43	121·4668	68	192·0871	93	262·7073
19	53·6714	44	124·2916	69	194·9119	94	265·5321
20	56·4962	45	127·1165	70	197·7367	95	268·3570
21	59·3210	46	129·9413	71	200·5615	96	271·1818
22	62·1458	47	132·7661	72	203·3863	97	274·0066
23	64·9706	48	135·5909	73	206·2111	98	276·8314
24	67·7954	49	138·4157	74	209·0359	99	279·6562
25	70·6203	50	141·2405	75	211·8608	100	282·4810

Miles per US gallon
to kilometres per litre

miles/ US gal	km/l	miles/ US gal	km/l	miles/ US gal	km/l	miles/ US gal	km/l
1	0·4251	26	11·0537	51	21·6823	76	32·3109
2	0·8503	27	11·4789	52	22·1075	77	32·7361
3	1·2754	28	11·9040	53	22·5326	78	33·1612
4	1·7006	29	12·3292	54	22·9578	79	33·5864
5	2·1257	30	12·7543	55	23·3829	80	34·0115
6	2·5509	31	13·1795	56	23·8081	81	34·4367
7	2·9760	32	13·6046	57	24·2332	82	34·8618
8	3·4012	33	14·0298	58	24·6584	83	35·2870
9	3·8263	34	14·4549	59	25·0835	84	35·7121
10	4·2514	35	14·8800	60	25·5086	85	36·1372
11	4·6766	36	15·3052	61	25·9338	86	36·5624
12	5·1017	37	15·7303	62	26·3589	87	36·9875
13	5·5269	38	16·1555	63	26·7841	88	37·4127
14	5·9520	39	16·5806	64	27·2092	89	37·8378
15	6·3772	40	17·0058	65	27·6344	90	38·2630
16	6·8023	41	17·4309	66	28·0595	91	38·6881
17	7·2274	42	17·8560	67	28·4846	92	39·1132
18	7·6526	43	18·2812	68	28·9098	93	39·5384
19	8·0777	44	18·7063	69	29·3349	94	39·9635
20	8·5029	45	19·1315	70	29·7601	95	40·3887
21	8·9280	46	19·5566	71	30·1852	96	40·8138
22	9·3532	47	19·9818	72	30·6104	97	41·2390
23	9·7783	48	20·4069	73	31·0355	98	41·6641
24	10·2035	49	20·8321	74	31·4607	99	42·0893
25	10·6286	50	21·2572	75	31·8858	100	42·5144

Kilometres per litre
to miles per US gallon

km/l	miles/ US gal	km/l	miles/ US gal	km/l	miles/ US gal	km/l	miles US gal
1	2·3522	26	61·1559	51	119·9597	76	178·7634
2	4·7043	27	63·5081	52	122·3118	77	181·1156
3	7·0565	28	65·8602	53	124·6640	78	183·4677
4	9·4086	29	68·2124	54	127·0161	79	185·8199
5	11·7608	30	70·5645	55	129·3683	80	188·1720
6	14·1129	31	72·9167	56	131·7204	81	190·5242
7	16·4651	32	75·2688	57	134·0726	82	192·8763
8	18·8172	33	77·6210	58	136·4247	83	195·2285
9	21·1694	34	79·9731	59	138·7769	84	197·5806
10	23·5215	35	82·3253	60	141·1290	85	199·9328
11	25·8737	36	84·6774	61	143·4812	86	202·2849
12	28·2258	37	87·0296	62	145·8333	87	204·6371
13	30·5780	38	89·3817	63	148·1855	88	206·9892
14	32·9301	39	91·7339	64	150·5376	89	209·3414
15	35·2823	40	94·0860	65	152·8898	90	211·6935
16	37·6344	41	96·4382	66	155·2419	91	214·0457
17	39·9866	42	98·7903	67	157·5941	92	216·3978
18	42·3387	43	101·1425	68	159·9462	93	218·7500
19	44·6909	44	103·4946	69	162·2984	94	221·1021
20	47·0430	45	105·8468	70	164·6505	95	223·4543
21	49·3952	46	108·1989	71	167·0027	96	225·8064
22	51·7473	47	110·5511	72	169·3548	97	228·1586
23	54·0995	48	112·9032	73	171·7070	98	230·5107
24	56·4516	49	115·2554	74	174·0591	99	232·8629
25	58·8038	50	117·6075	75	176·4113	100	235·2150

Degrees Celsius
to degrees Fahrenheit

°C	°F	°C	°F	°C	°F	°C	°F
− 50	− 58·0	− 25	− 13·0	0	32·0	25	77·0
− 49	− 56·2	− 24	− 11·2	1	33·8	26	78·8
− 48	− 54·4	− 23	− 9·4	2	35·6	27	80·6
− 47	− 52·6	− 22	− 7·6	3	37·4	28	82·4
− 46	− 50·8	− 21	− 5·8	4	39·2	29	84·2
− 45	− 49·0	− 20	− 4·0	5	41·0	30	86·0
− 44	− 47·2	− 19	− 2·2	6	42·8	31	87·8
− 43	− 45·4	− 18	− 0·4	7	44·6	32	89·6
− 42	− 43·6	− 17	1·4	8	46·4	33	91·4
− 41	− 41·8	− 16	3·2	9	48·2	34	93·2
− 40	− 40·0	− 15	5·0	10	50·0	35	95·0
− 39	− 38·2	− 14	6·8	11	51·8	36	96·8
− 38	− 36·4	− 13	8·6	12	53·6	37	98·6
− 37	− 34·6	− 12	10·4	13	55·4	38	100·4
− 36	− 32·8	− 11	12·2	14	57·2	39	102·2
− 35	− 31·0	− 10	14·0	15	59·0	40	104·0
− 34	− 29·2	− 9	15·8	16	60·8	41	105·8
− 33	− 27·4	− 8	17·6	17	62·6	42	107·6
− 32	− 25·6	− 7	19·4	18	64·4	43	109·4
− 31	− 23·8	− 6	21·2	19	66·2	44	111·2
− 30	− 22·0	− 5	23·0	20	68·0	45	113·0
− 29	− 20·2	− 4	24·8	21	69·8	46	114·8
− 28	− 18·4	− 3	26·6	22	71·6	47	116·6
− 27	− 16·6	− 2	28·4	23	73·4	48	118·4
− 26	− 14·8	− 1	30·2	24	75·2	49	120·2

Degrees Celsius to degrees Fahrenheit

°C	°F	°C	°F	°C	°F	°C	°F
50	122·0	75	167·0	100	212·0	125	257·0
51	123·8	76	168·8	101	213·8	150	302·0
52	125·6	77	170·6	102	215·6	175	347·0
53	127·4	78	172·4	103	217·4	200	392·0
54	129·2	79	174·2	104	219·2	225	437·0
55	131·0	80	176·0	105	221·0	250	482·0
56	132·8	81	177·8	106	222·8	275	527·0
57	134·6	82	179·6	107	224·6	300	572·0
58	136·4	83	181·4	108	226·4	325	617·0
59	138·2	84	183·2	109	228·2	350	662·0
60	140·0	85	185·0	110	230·0	375	707·0
61	141·8	86	186·8	111	231·8	400	752·0
62	143·6	87	188·6	112	233·6	425	797·0
63	145·4	88	190·4	113	235·4	450	842·0
64	147·2	89	192·2	114	237·2	475	887·0
65	149·0	90	194·0	115	239·0	500	932·0
66	150·8	91	195·8	116	240·8	525	977·0
67	152·6	92	197·6	117	242·6	550	1022·0
68	154·4	93	199·4	118	244·4	575	1067·0
69	156·2	94	201·2	119	246·2	600	1112·0
70	158·0	95	203·0	120	248·0	625	1157·0
71	159·8	96	204·8	121	249·8	650	1202·0
72	161·6	97	206·6	122	251·6	675	1247·0
73	163·4	98	208·4	123	253·4	700	1292·0
74	165·2	99	210·2	124	255·2	1000	1832·0

Degrees Fahrenheit to degrees Celsius

°F	°C	°F	°C	°F	°C	°F	°C
− 50	− 45·6	− 25	− 31·7	0	− 17·8	25	− 3·9
− 49	− 45·0	− 24	− 31·1	1	− 17·2	26	− 3·3
− 48	− 44·4	− 23	− 30·6	2	− 16·7	27	− 2·8
− 47	− 43·9	− 22	− 30·0	3	− 16·1	28	− 2·2
− 46	− 43·3	− 21	− 29·4	4	− 15·6	29	− 1·7
− 45	− 42·8	− 20	− 28·9	5	− 15·0	30	− 1·1
− 44	− 42·2	− 19	− 28·3	6	− 14·4	31	− 0·6
− 43	− 41·7	− 18	− 27·8	7	− 13·9	32	0·0
− 42	− 41·1	− 17	− 27·2	8	− 13·3	33	0·6
− 41	− 40·6	− 16	− 26·7	9	− 12·8	34	1·1
− 40	− 40·0	− 15	− 26·1	10	− 12·2	35	1·7
− 39	− 39·4	− 14	− 25·6	11	− 11·7	36	2·2
− 38	− 38·9	− 13	− 25·0	12	− 11·1	37	2·8
− 37	− 38·3	− 12	− 24·4	13	− 10·6	38	3·3
− 36	− 37·8	− 11	− 23·9	14	− 10·0	39	3·9
− 35	− 37·2	− 10	− 23·3	15	− 9·4	40	4·4
− 34	− 36·7	− 9	− 22·8	16	− 8·9	41	5·0
− 33	− 36·1	− 8	− 22·2	17	− 8·3	42	5·6
− 32	− 35·6	− 7	− 21·7	18	− 7·8	43	6·1
− 31	− 35·0	− 6	− 21·1	19	− 7·2	44	6·7
− 30	− 34·4	− 5	− 20·6	20	− 6·7	45	7·2
− 29	− 33·9	− 4	− 20·0	21	− 6·1	46	7·8
− 28	− 33·3	− 3	− 19·4	22	− 5·6	47	8·3
− 27	− 32·8	− 2	− 18·9	23	− 5·0	48	8·9
− 26	− 32·2	− 1	− 18·3	24	− 4·4	49	9·4

Degrees Fahrenheit to degrees Celsius

°F	°C	°F	°C	°F	°C	°F	°C
50	10·0	75	23·9	100	37·8	125	51·7
51	10·6	76	24·4	101	38·3	126	52·2
52	11·1	77	25·0	102	38·9	127	52·8
53	11·7	78	25·6	103	39·4	128	53·3
54	12·2	79	26·1	104	40·0	129	53·9
55	12·8	80	26·7	105	40·6	130	54·4
56	13·3	81	27·2	106	41·1	131	55·0
57	13·9	82	27·8	107	41·7	132	55·6
58	14·4	83	28·3	108	42·2	133	56·1
59	15·0	84	28·9	109	42·8	134	56·7
60	15·6	85	29·4	110	43·3	135	57·2
61	16·1	86	30·0	111	43·9	136	57·8
62	16·7	87	30·6	112	44·4	137	58·3
63	17·2	88	31·1	113	45·0	138	58·9
64	17·8	89	31·7	114	45·6	139	59·4
65	18·3	90	32·2	115	46·1	140	60·0
66	18·9	91	32·8	116	46·7	141	60·6
67	19·4	92	33·3	117	47·2	142	61·1
68	20·0	93	33·9	118	47·8	143	61·7
69	20·6	94	34·4	119	48·3	144	62·2
70	21·1	95	35·0	120	48·9	145	62·8
71	21·7	96	35·6	121	49·4	146	63·3
72	22·2	97	36·1	122	50·0	147	63·9
73	22·8	98	36·7	123	50·6	148	64·4
74	23·3	99	37·2	124	51·1	149	65·0

Degrees Fahrenheit
to degrees Celsius

°F	°C	°F	°C	°F	°C	°F	°C
150	65·6	175	79·4	200	93·3	225	107·2
151	66·1	176	80·0	201	93·9	250	121·1
152	66·7	177	80·6	202	94·4	275	135·0
153	67·2	178	81·1	203	95·0	300	148·9
154	67·8	179	81·7	204	95·6	325	162·8
155	68·3	180	82·2	205	96·1	350	176·7
156	68·9	181	82·8	206	96·7	375	190·6
157	69·4	182	83·3	207	97·2	400	204·4
158	70·0	183	83·9	208	97·8	425	218·3
159	70·6	184	84·4	209	98·3	450	232·2
160	71·1	185	85·0	210	98·9	475	246·1
161	71·7	186	85·6	211	99·4	500	260·0
162	72·2	187	86·1	212	100·0	525	273·9
163	72·8	188	86·7	213	100·6	550	287·8
164	73·3	189	87·2	214	101·1	575	301·7
165	73·9	190	87·8	215	101·7	600	315·6
166	74·4	191	88·3	216	102·2	625	329·4
167	75·0	192	88·9	217	102·8	650	343·3
168	75·6	193	89·4	218	103·3	675	357·2
169	76·1	194	90·0	219	103·9	700	371·1
170	76·7	195	90·6	220	104·4	725	385·0
171	77·2	196	91·1	221	105·0	750	398·9
172	77·8	197	91·7	222	105·6	775	412·8
173	78·3	198	92·2	223	106·1	800	426·7
174	78·9	199	92·8	224	106·7	1000	537·8